Make:

Radio

HANDS-ON ADVENTURES IN THE HIDDEN UNIVERSE OF RADIO WAVES

Fredrik Jansson

Make:
RADIO

By Fredrik Jansson

ISBN: 978-1-68045-677-6

July 2024: First Edition

See www.oreilly.com/catalog/errata.csp?isbn=9781680456776 for release details.

Make: BOOKS
President Dale Dougherty
Creative Director Juliann Brown
Editor Kevin Toyama
Illustrator Charles Platt
Designer SeeSullivan
Copyeditor Sophia Smith
Proofreader Mark Nichol

Make Community, LLC
150 Todd Road, Suite 100
Santa Rosa, California 95407

www.make.co

CONTENTS:

Experiment 1

Experiment 2

Experiment 3

Experiment 4

CONTENTS:

Experiment 8

Experiment 9

Experiment 10

CONTENTS:

CONTENTS:

FOREWORD BY CHARLES PLATT

THE MYSTERY AND MAGIC OF RADIO

Radio waves are all around us, transmitting a fantastic variety of data on thousands of different frequencies. Whenever a piece of electronic equipment communicates via electromagnetic radiation, it is a form of radio:

- Your cell phone emits a radio signal when you speak into it.
- Emergency responders exchange radio messages via walkie-talkies.
- An Ethernet router that uses Wi-Fi to connect with your computer is a form of radio.
- Bluetooth is a radio link between devices such as a phone and wireless headphones.
- If you have a little plastic box in your car that opens your garage door when you press a button, the box contains a radio transmitter.
- If your home has a TV dish, you are receiving radio transmissions from a satellite that is probably 22,000 miles away.
- Low-orbit satellites such as the Starlink network are interacting with ground antennas via radio waves.

Somehow, radio waves convey power from a transmitter to a receiver. We take this for granted, but there's a feeling of mystery about information traveling through the air without needing electrons to carry it.

This book will help you to understand the phenomenon as you build your own transmitters and receivers using a limited number of affordable components. You'll learn how audio is added to a carrier wave, how FM radio encodes its signals in a very different way from AM radio, and how a microcontroller can be used to transmit radio while also showing you information about the signals you receive.

F-1 *Western Union messenger boys before the telephone was invented.*

To gain a better appreciation for the modern world of radio, it's useful to imagine how the world used to be without it.

SENDING INFORMATION

In 1800, the only way to deliver a handwritten message over a long distance in the United States was by writing a letter and sending it via the US Postal Service. Moreover, since motor vehicles had not been invented and railroads in the United States had not been built from coast to coast, letters that crossed the nation were carried for some of the distance by horse-drawn vehicles or by horseback. If you were living in New York City, and you wanted to know what was happening to relatives who had relocated to California, you might have to wait a couple of months to find out.

During the 1840s, telegraph systems were developed, typically using a single wire strung alongside a railroad, with the steel track completing the circuit. The wire carried electrical signals in Morse code—the system of short and long pulses invented by Samuel Morse. A telegraph operator sent the pulses manually by tapping a lever that closed a pair of contacts. At the receiving end, an electromagnet pulled a stylus that made impressions in a paper tape. An operator then transcribed the code to English in longhand, and the transcript was delivered as a telegram by a messenger boy, often one riding a bicycle. Figure **F-1** shows some messenger boys of the period.

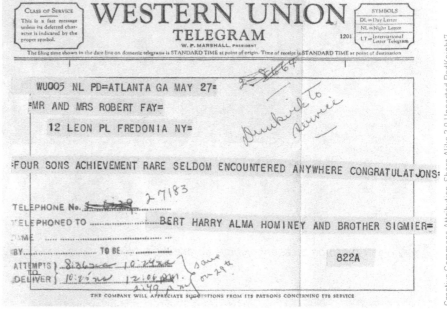

F-2 *A Western Union Telegram (USA), including annotations regarding attempted delivery. The message offers congratulations on the birth of a fourth son. Although sent in 1959, the format had been established decades earlier.*

The first transcontinental telegraph line was established in the United States in 1861. By this time, the printing telegraph had been developed, and telegrams evolved to a format that remained fixed for many decades. Text was printed on little paper tapes and pasted to a piece of yellow backing paper, like the example in Figure **F-2**.

While the task of communicating via telegram seems primitive and laborious to us now, it created an ability to send important news rapidly that seemed miraculous at the time.

In the 1880s, an even more impressive miracle appeared: telephone service. Now, you could actually hear someone's voice at a distance—if the distance was not too great and you listened very carefully. The sounds were faint because amplifiers had not yet been developed.

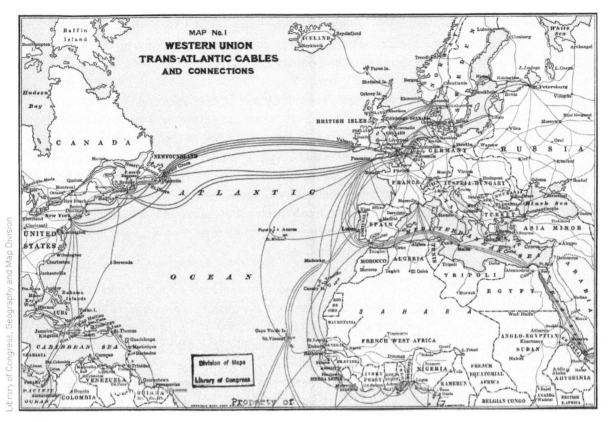

F-3 *Undersea transatlantic cables in 1900.*

One major obstacle still remained: the oceans of the world. Telegraph cables had been laid across the Atlantic, but they were still limited to Morse code, and they served fixed stations on land. (See Figure **F-3**.) Sending a message from ship to ship or shore to shore required flashing a signal lamp while the person receiving the message watched through a telescope. The maximum range was only about five miles, as the curvature of the Earth meant that a distant ship was below the horizon. If an admiral on land wanted to send new orders to a captain at sea, he couldn't.

THE DAWN OF RADIO

In 1895, Guglielmo Marconi was a university student studying physics in Italy. When his tutor mentioned that Heinrich Hertz had proved the existence of electromagnetic radiation by generating electric sparks and detecting them at the opposite side of a room, Marconi was amazed that no one had tried to develop this concept to make money out of it. He imagined that electrical communication through the air could be as important, and as lucrative, as sending messages through wires.

He was not an especially gifted scientist, but had grown up in a wealthy family that gave him a sophisticated understanding of money. He was also enterprising, innovative, and persistent, and he started experimenting. Quickly, he discovered that longer antenna wires on the transmitter and on the receiver would greatly extend their range. He also found that the range increased if he grounded the transmitter and the receiver. The messages that he transmitted consisted only of little bursts of electrical noise, but they would be good enough to enable Morse code. Amplifiers still didn't exist, so headphones were necessary.

At the age of 22, Marconi conducted demonstrations in London, trying to attract interest and funding from the military. A year later, he managed to transmit a signal over a distance of four miles, using antennas carried by balloons. His efforts attracted publicity, and the publicity attracted investment capital, which he used to refine his equipment. By 1900, he was able to send a signal between two ships that were almost 60 miles apart.

Shrewdly, he added prestige to his enterprise by retaining some of the world's most impressive scientists, including Lord Kelvin, who had formulated the first two laws of thermodynamics, and John Ambrose Fleming, who later invented the first thermionic vacuum tube.

Marconi's ultimate goal was to send a signal across the Atlantic Ocean between southwest England and Massachusetts. This would be his most impressive proof-of-concept demo, and he took the risk of publicizing it ahead of time.

Huge transmitting towers were erected but were wrecked by a high wind. After other setbacks, while the world waited to see if Marconi could fulfill his predictions, he set up simpler antennas to send a prearranged signal over a slightly shorter distance, to Newfoundland, where he established a receiving station. On two designated days, he and an assistant spent hours listening on headphones in an effort to detect clicking sounds amid the radio static. Twice, they claimed they heard the signal.

Scientists were skeptical because radio waves, like light waves, should be blocked by the curvature of the Earth. However, electrical engineer A. E. Kennelly and physicist Oliver Heaviside, working independently, theorized that an ionized layer in the atmosphere might bounce radio signals around the surface of the planet. Marconi liked their theory, as it validated his claims. The existence of the Heaviside layer, as it is now known, was not confirmed until many years later.

However, the publicity generated by Marconi's transatlantic experiment enabled him to raise more money. He patented all the key concepts, and the company he established became the dominant international manufacturer of wireless stations. Initially, his equipment was used mostly for military and marine messages because everyday people had no special interest in listening to the beeps of Morse code. But a brilliant electrical engineer named E. Howard Armstrong came up with ideas that improved the quality of radio reception to the point where you could transmit and receive speech and music, at which time the real radio revolution began.

First, Armstrong invented the regenerative circuit, which used positive feedback through a vacuum tube, enabling people to listen to radio using a loudspeaker instead of headphones. Armstrong filed a patent describing the concept in 1913, then came up with an idea for a circuit that he named the superheterodyne, which had better selectivity and stability and created less noise. He filed for a patent for that in 1919.

Radio-frequency spectrum was allocated internationally in 1912, but any enthusiast could set up a transmitter, and some people used crystal sets to listen in to the airwaves just to see what they could find. Then in 1934, the US federal government enacted the Communications Act, which licensed companies to transmit signals without the risk of interfering with each other.

This enabled a new era of broadcasting. Radio was no longer limited to communication between two points; transmissions radiated from a radio station to thousands or even millions of listeners who received it simultaneously. Funded by advertising in the United States (and by government money in many other nations), radio brought big band music, comedians, dramas, sports broadcasts, political speeches, and newscasters into your living room. Families would gather attentively around the radio

F-4 *A tabletop radio receiver in real wood veneer, typical in those days.*

F-5 *In the 1930s, a receiver such as this could occupy a dominant position in a living room.*

receiver, listening to distant stations that carried their favorite shows.

Early radio receivers were enshrined in fancy cabinets, suggesting the importance and magic associated with voices and music traveling through the air. Examples are shown in Figures **F-4** and **F-5**.

Radio became the primary medium of entertainment in America, and rivaled newspapers as a source of information. It maintained that status until television triggered another revolution, and the era when millions of people sat listening to comedy or drama gradually came to an end.

Decades passed before radio was readopted for short-range purposes, such as unlocking your car when you press a button on a remote or transmitting music from your phone to a pair of earbuds.

This book will not only show you how to pick up signals from AM and FM stations—which still broadcast music and voices around the world—but will also teach you to transmit your own radio signals. Eventually you'll be able to acquire your own shortwave license if you choose.

I think you will find that it is an exciting journey.

SOURCES

- www.elon.edu/u/imagining/time-capsule/150-years/back-1830-1860/
- www.historyofinformation.com/detail.php?entryid=580
- lemelson.mit.edu/resources/samuel-morse
- www.loc.gov/resource/g9101p.ct003867/?r=0.103,0.116,0.777,0.577,0
- *Much Ado About Almost Nothing*,by Hans Camenzind, a history of electronics, republished in 2023 by Faraday Press
- museumsanfernandovalley.blogspot.com/2008/10/butterfield-stage-on-valleys-el-camino.html
- www.smithsonianmag.com/smithsonian-institution/brief-history-united-states-postal-service-180975627/
- som.csudh.edu/cis/471/hout/telecomhistory/
- www.telegraph-history.org/transcontinental-telegraph/
- truewestmagazine.com/an-in-depth-look-at-the-telegraph-system-in-the-old-west/

PREFACE

As a kid, I used to take apart radios and electronic devices. I found circuit boards with mysterious small things attached, or—if the radio was very old—electron tubes made of glass with tiny metal structures inside. I wanted to know how these worked and build my own electronic circuits, and this got me reading about electronics in books in the local library.

Radio, in particular, is fascinating for me, probably because it combines the abstract physics of radio waves (electromagnetism) and practical tinkering with electronics. I realized that by assembling the right circuit, I should be able to generate or detect radio waves, which meant I would be able to send or receive messages over a distance.

In practice, however, this radio building turned out to be difficult. I mostly found electronics books for radio amateurs about building transmitters and receivers: the problem is that transmitting on amateur frequencies requires a license. Often, they also used components I couldn't get hold of because the components were obsolete or because the local electronics store didn't have them in stock. Nowadays, components are much more accessible, as it is possible to order them online from a variety of stores.

My goal for this book is to provide radio experiments that can be built with obtainable components and that can be put in use without needing a license. I've tried to keep the circuits simple (sometimes at the expense of features or performance) with the thought that one learns more from actually building a simple radio circuit than from just planning a complicated one.

In the end, I did obtain an amateur radio license, and I have had a lot of fun with also that aspect of radio. The final chapter of the book gives some pointers for how to get licensed, along with other ways to continue with radio as a hobby.

INTRODUCTION

BEFORE YOU BEGIN

The purpose of this introduction is to provide some quick orientation regarding the knowledge that you need and the components that you will be using for hands-on work throughout the book. If you need more detailed information about components, tools, supplies, and basic concepts, you can find it in Appendices A, B, C, and D.

I hope that you will already understand a few basics of electronics, such as these topics:

- How to insert components into a breadboard.
- How to make jumpers for your breadboard by cutting hookup wire and stripping insulation off the ends.
- What resistors are for.
- What an integrated circuit chip looks like.
- How to use a multimeter to measure voltage, current, resistance, and (ideally) capacitance.

If you are not familiar with these concepts, you will still be able to build the circuits in this book by following my instructions. But your learning experience will be more meaningful if you understand what you are doing, and for this purpose, I recommend an intro-level guide such as *Make: Electronics* by Charles Platt.

THE LEARNING PROCESS

I like the process of learning by discovery. This means that instead of me telling you how everything works, you perform experiments and draw your own conclusions as much as possible. The hands-on experience of putting components together and seeing what happens is the best way to learn, besides being a lot of fun. With this in mind, I am hoping you will build the circuits in this book, using components that are listed in Appendix A.

Appendix B lists some suggested sources for these components. If you prefer not to shop around, kits have been developed specifically for the projects in this book. You will find them at ProTechTrader.com.

All of my circuits can be assembled using one or two solderless breadboards. However, if you want to keep a circuit for permanent use, you may wish to make soldered connections. I don't have space to teach you about soldering, but many guides are available if you want to try it.

PREPARATION

The projects in this book will require you to have a basic setup on a workbench or table. The setup will include some tools and supplies:

- A solderless breadboard.
- A power supply, either a 9V battery or an AC adapter with a 9VDC output. (For some projects, you must use a 9V battery or a well-regulated benchtop power supply because an AC adapter is likely to create radio interference.)
- 22-gauge solid hookup wire in red, green, yellow, and blue or black.
- A multimeter.
- Long-nose pliers, wire cutters, and wire strippers.
- A set of miniature screwdrivers.
- Assorted jumper wires. You can make them yourself or buy them premade.

SCHEMATICS, DIAGRAMS, AND PHOTOGRAPHS

You will find three types of illustrations in this book.

Photographs show you what components look like and what a finished circuit should look like. They have limitations, though, because you can't always see the values of the components, and some will be partially hidden by others. Therefore, I will include pictorial breadboard diagrams using the symbols shown in Figure **0-1**. Note that wherever you see little pink dots, they tell you the locations of pins that are hidden under a component.

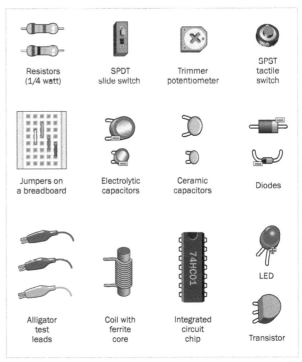

0-1 *Pictorial representations that are used throughout the book.*

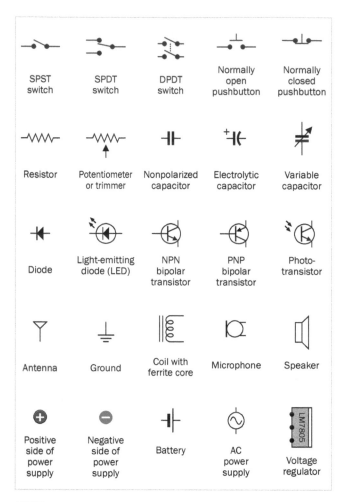

0-2 *The meaning of basic schematic symbols.*

In pictorial diagrams, try to remember my use of color for jumper wires:

- Red wires are connected with the positive power supply.
- Blue wires are connected with negative ground.
- Other wires are green or yellow.

When I want to explain how a circuit works, I use schematics with symbols such as those shown in Figure **0-2**.

Also, remember these rules:

- Where a wire in a circuit makes an electrical connection with another wire, they are always connected with a dot.
- If a wire crosses another wire without a dot, there is no connection between them.

If you look at schematic diagrams in other books or online, they usually show the positive power supply at the top and negative ground at the bottom, while an input to the circuit will be on the left and an output will be on the right. This is especially true where schematics of radio receivers are involved. I have chosen to follow this convention because it is almost universal. The problem is that if you build a circuit using a breadboard in its normal, vertical orientation, power is delivered along the sides of the board, while a signal usually flows from top to bottom. You will find yourself faced with a tricky mental exercise if you look at a radio schematic and try to convert it to a breadboard layout.

Bearing this in mind, the breadboards in this book are depicted horizontally. You can always rotate the book by 90 degrees if you want to view a breadboard in its more usual orientation.

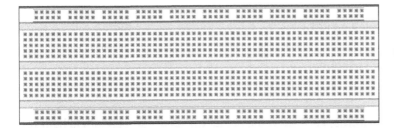

0-3 *I will use this type of breadboard throughout the book.*

0-4 *The conductors that are hidden inside a breadboard, connecting components that are inserted into the holes.*

Figure **0-3** shows the standard type of breadboard that I will use, with 830 holes, properly known as tie points. The two pairs of long, horizontal rows of holes with colored stripes beside them are called buses, and we will try to use just the top, positive bus and bottom, negative bus, in accordance with our plan to have conventional current flowing from top to bottom.

Some breadboards have a break midway in each bus, requiring you to bridge those gaps with jumper wires. These boards have become uncommon, but if you want to make sure that your breadboard has unbroken buses, you can insert a piece of hookup wire at each end of a bus, set your meter to measure continuity, and test to make sure that the bus is continuous from end to end. If it isn't, add jumpers to bridge the gaps in the buses.

Figure **0-4** shows the configuration of metal conductors that are hidden inside a breadboard. These conductors connect the tie points in each row and each bus.

In each schematic, components will be identified with abbreviations such as *R1* and *R2* (for resistors) or *C1* and *C2* (for capacitors) so that we can refer to the parts easily. The abbreviations for commonly used components are as follows:

- *R:* resistor
- *P:* potentiometer or trimmer
- *C:* capacitor
- *IC:* integrated circuit chip

- **Q:** transistor
- **D:** diode (including LED)
- **L:** coil
- **S:** switch or pushbutton
- **LS:** loudspeaker

In pictorial diagrams, I will show the actual values of components to help you when you are building the circuit yourself.

IF SOMETHING DOESN'T WORK

Usually, a circuit will only work if you build it without making any mistakes. Unfortunately, all of us do make mistakes, so the odds are against you if you don't proceed in a methodical manner.

I know how frustrating it is when components just sit there doing nothing, but if you build a circuit that doesn't work, getting annoyed with it will actually make you less likely to see the fault. The best way to find the problem is to be patient and examine every detail systematically.
These are the most likely issues:

- You made a wiring error. This happens to everyone, including me. Your chances of seeing the error will improve if you walk away from your work table for half an hour and do something else before returning to take another look.
- You may have overloaded a component such as a transistor or a chip so that it doesn't work anymore. Try to keep some spares just in case. Learn how to test transistors with your meter.
- You may have inserted the wrong component. For instance, a 100-ohm resistor is easily confused with a 1K resistor, as only one stripe has a different color. You can save time later if you devote a little time initially to testing each component with your meter before inserting it in a board.
- There may be a bad connection between a component and the metal strip inside a breadboard. Try wiggling loose components, measuring voltages, and, if necessary, moving key components to a slightly different location on the board.
- You may have inserted a component in a row of holes adjacent to the correct row. This is a very common error.

If you get desperate, you can email me to ask for help—but please try everything else first, and please be patient while waiting for a reply.

Sometimes, you'll receive a reply on the same day, especially if you're reporting an error that I have made. Other times, you may have to wait for several days. Remember the following tips:

- Attach photographs of any project that doesn't work. I must be able to see details such as the colors of stripes on resistors.
- Clearly identify which project you have been working on, and which book you are referring to. Include the figure number of any schematic or photograph that you mention.
- Describe the problem clearly, in the same style as if you were describing a physical symptom to a doctor and asking for a diagnosis.

Send your message to fjansson@abo.fi and put *HELP* in the subject line.

REPORTING A MISTAKE

No matter how many times I read the text in this book and check the illustrations, I will miss some little errors. If you find one, please report it. You can use my email address for that purpose, or you can go to the Errata page maintained by O'Reilly and Associates, which distributes this book. The advantage of sending email is that we can discuss the problem if necessary. The advantage of the O'Reilly system is that you can read other people's reports and see if your issue has already been resolved. Also, after you make a report to the O'Reilly website, other people can read it. The O'Reilly site's errata page for this book is at: www.oreilly.com/catalog/errata.csp?isbn=9781680456776.

GOING PUBLIC

If you run into some kind of problem, you may want to complain about it. One way in which people complain is in reader reviews online. Naturally, I respect your right to express yourself, but if you are annoyed about something, please contact me first to give me a chance to resolve your complaint.

Be aware of the power that you have as a reader, and please use it fairly. A single negative review can create a bigger effect than you may realize. It can outweigh a half dozen positive reviews. My star ratings are very valuable to me!

1 POWER THROUGH THIN AIR

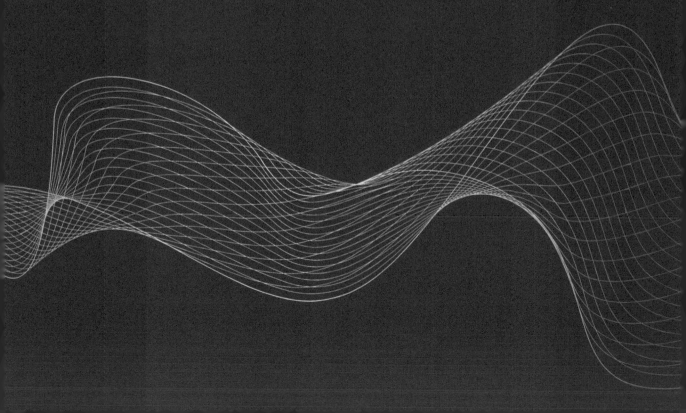

Maybe you think of a radio transmitter as being a tower 200 feet high, standing out in a field somewhere, or a heavy-duty military device on a combat vehicle. But a transmitter doesn't have to be large, complicated, or expensive. You can use just a dozen cheap components on your kitchen table to deliver a radio transmission.

In this first experiment, two basic timer chips will create a *signal* and a *carrier wave*, which are two of the most important concepts in radio. After you apply power to the circuit, you can tune in to its transmission using even the cheapest handheld AM radio, or you can build your own tiny receiver, which requires only five components.

Along the way, I will explain the difference between frequency and wavelength. I will show you a variable capacitor, an inductor, and a diode, and you will wrap wire around a ferrite rod to form a compact, versatile antenna.

You won't be doing any voice transmissions just yet, but you will learn about the fundamental ideas that form a basis for radio.

ALL ABOUT AUDIO

The concepts of *frequency* and *wavelength* are fundamental in radio, so I'll begin with a quick demo to clear up any misconceptions.

You Will Need:

- Snap-on connector for your 9V battery or alligator leads to grip the terminals (1).
- 9V battery (1).
- SPDT switch with pins spaced 1/10" for insertion in breadboard (1).
- 7555 integrated circuit chips (2).
- European-style connection block with 12 pairs of terminals spaced 8mm or 5/16" (1).
- 22-gauge wire for jumpers (red, green, yellow, and blue or black; 10" of each color).
- 22-gauge wire for antenna and ground (a total of 40 feet, any color; you can join shorter pieces, unjoin them. and use them later).
- 26-gauge wire for coil (4 feet). The copper inside the insulation should be solid, not stranded.
- Ferrite rod, 3/8" diameter, 6" long (1). A longer rod is acceptable. A slightly shorter rod may work but will be less effective.
- Variable capacitor, 200pF, type 223P (1). May also be described as a tuning capacitor. A plastic tuning wheel should be available, and should be added.
- BAT48 Schottky diode (1).
- Resistors: 100 ohms (1), 330 ohms (1), 2.2K (2), 10K (2).
- Trimmer potentiometer, 500K (1). Pin configurations of trimmers may vary, and must fit your breadboard; check Appendix A before ordering.
- Ceramic capacitors: 10pF (1), 100pF (1), 10nF (3), 0.1μF (1).
- Electrolytic capacitor, 100μF (1).
- Speaker, between 2" and 3" diameter (1). Earphone, must be high-impedance type, often sold for crystal radios (1). Alternatively, passive piezo speaker (1). See Appendix A for details.
- Small piece of card stock, such as index card.
- Clear adhesive tape.
- Optional: Project box in which to mount the speaker (1).
- Optional: Block of wood (1), about 4" x 4" x 2", and ¾" #2 pan-head screws (2).
- Optional: Cheapest possible handheld transistor radio, to test your signal (1). It must receive AM broadcasts in the medium-wave band. The letters *AM* should appear on the tuning dial of the radio, with numbers at intervals from 530 to slightly below 1700.

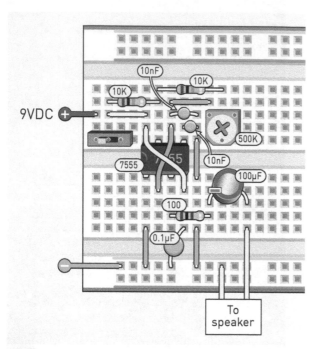

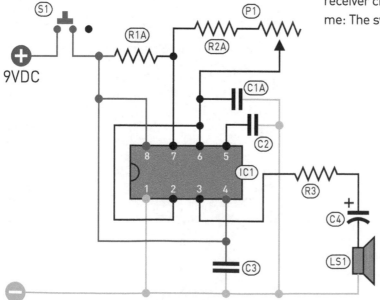

Remember that throughout this book, I will show breadboards horizontally because current in schematics is usually shown flowing from top to bottom, while signals in amplifiers and radios are usually shown flowing from left to right.

Figure **1-1** shows an audio test circuit built around a 7555 timer. This is a very common circuit, but we'll be using it in an unusual way. Everything is squeezed in fairly tightly, so count the holes in the board carefully when you insert the components. (If you are wondering why we are not using the positive bus at the top of the board, you'll see that we do use it later.)

Figure **1-2** shows the schematic version. Note that S1 functions as a power switch; nothing will happen until you slide the switch to the left. The need for this switch will become apparent later, when you add a receiver circuit to the same breadboard. Trust me: The switch is necessary!

1-1 *A basic circuit to generate audio frequencies.*

1-2 *The schematic version of the breadboard circuit from Figure 1-1.*

Components	
S1	Slide switch, SPDT
R1A	Timing resistor, 10K
R2A	Timing resistor, 10K
R3	Current limiting, 100 ohms
C1A	Timing capacitor, 10nF
C2	Bypass capacitor, 10nF
C3	Bypass capacitor, 0.1µF
C4	DC blocking, 100µF
IC1	7555 timer chip
LS1	8-ohm speaker, 2" minimum

The 7555 (labeled IC1 in the schematic) is a slightly newer version of the old 555 timer, which is the most widely used and longest-lived chip ever made. The original 555 is not appropriate here, because it generates a noisy output and does not run fast enough for the extension of the circuit we will be adding. The specification for 7555 chips varies slightly from one manufacturer to another, but so far as I know, all 7555 chips will work in the circuits in this experiment.

In Figure 1-1, the square blue object with a white circle and a cross on it is a 500K trimmer potentiometer—that is, a variable resistance that you can adjust by turning a screw in the top of it so that its resistance varies from 0 ohms to 500,000 ohms. The potentiometer is identified as P1 in Figure 1-2.

Because I'll be referring to some of the pins on the 7555 timer by name, I have summarized them in Figure **1-3**. I will also refer to pins by number, so remember that chips always number their pins counterclockwise, seen from above, beginning from the top-left corner, when the semicircular notch of the chip is at the top.

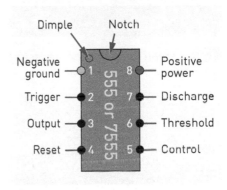

1-3 *The names of pins on a 555 or 7555 timer chip.*

As soon as you apply 9VDC to the circuit and switch it on, the timer emits a stream of pulses from its output pin. In the circuit, the pulses pass through a 100-ohm resistor (to limit the current) and a 100μF electrolytic capacitor (which blocks DC) on their way to a speaker. The duration of each positive pulse is determined by three resistances: R1A, R2A, and P1. The gaps between pulses are determined by R2A and P1. The size of capacitor C1A also determines the duration of the pulses and the gaps between them. Higher-value resistances and/or higher capacitance will generate longer pulses and gaps, so the potentiometer will adjust the pulse stream.

If you want to know how to calculate the results of resistance and capacitor values when using a 555 or 7555 chip, type this term into the search engine of your choice:

```
555 timer calculator
```

1-4 *A small loudspeaker prepackaged in a 4" enclosure.*

1-5 *A 3" loudspeaker mounted in a project box.*

Many calculators are available online, and the values for a 555 timer will create the same frequencies as when you use a 7555 timer.

When you adjust the trimmer potentiometer, P1, you should hear tones that span most of the range of human hearing. Why do you need to do this? Because sound is often transmitted by radio, you need a source for testing purposes. Also, knowing how frequency and wavelength function as parts of an audio signal is essential in understanding radio waves.

Incidentally, some trimmer potentiometers are not well designed for use with breadboards. Their little pins have kinks in them, which may be difficult to insert fully, and the trimmer may tend to work its way out of the board. You can overcome this tendency by using pliers to straighten or twist the pins.

SOUND QUALITY

When you hear sound from a naked speaker sitting on your workbench, waves of air pressure are coming from the rear of the speaker as well as from the front of it. The waves from the front and from the back tend to cancel each other out, so that low frequencies are not reproduced very well and the overall loudness of the speaker is reduced.

You can improve the sound quality radically by mounting the speaker in an enclosure. Some small speakers are sold in boxes specifically made for this purpose, such as the one in Figure **1-4**, which I found in a hardware store, designed for use with a door bell. Alternatively, you can make one yourself. Search online for:

`electronics project box`

You can buy the cheapest project box that's big enough for your speaker. Even a small cardboard box will provide noticeable improvement in sound quality. Drill holes in the lid and mount the speaker inside using epoxy glue or small nuts and bolts, as shown in Figure **1-5**.

You'll need to attach wires to the terminals on your speaker, and soldering is the best way to do this, but you can just wrap several turns of wire around the speaker terminals and use pliers to squeeze the wires tightly into position. Alligator clips may also be used.

If you prefer something that looks a bit better, you can pay a few extra dollars for a box such as the one in Figure **1-6**. The pattern of quarter-inch holes in the lid was created with drawing software and printed onto paper, as in Figure **1-7**. This was taped under the lid, and an awl was used to prick through the paper, marking the center of each circle.

The best way to drill holes in soft plastic is with a Forstner bit, but if you only have regular drill bits, make small holes first and then enlarge them. You can use a countersink for this purpose. If you start right in with a large bit, it will tend to dig into the plastic and make a mess.

Putting a speaker in a box may seem an unnecessary distraction from the topic of radio, but many of the projects in this book have an audio output, and it will sound much better if you put your speaker in an enclosure.

1-6 *A better-looking speaker enclosure.*

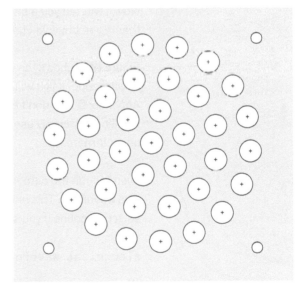

1-7 *A template for drilling holes in a box lid.*

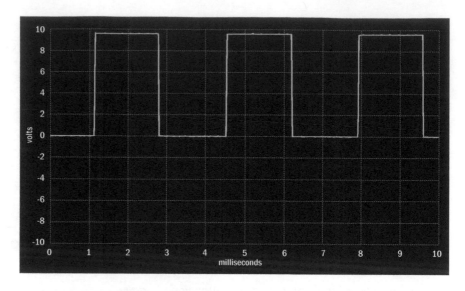

1-8 *A square wave output from a 7555 timer running in astable mode.*

FREQUENCY AND WAVELENGTH

Now, I will tell you a bit of theory about the audio output. You do not need this theory yet, but it will be necessary later.

Figure **1-8** shows the actual output displayed on the screen of an oscilloscope. If you want to know more about oscilloscopes, please turn to Appendix D. You don't have to own one to build and test the circuits here, but it will be extremely useful, and some oscilloscopes are now almost as cheap as multimeters.

Even though the output in Figure 1-8 consists of straight lines, it's known as a **waveform**. This one happens to be a square wave, but you can view a selection online if you search for

`electrical waveforms`

and view the search results as images. I'll be showing you radio waves before long, so it's a good idea to familiarize yourself with the concept.

The scale on the left in Figure 1-8 shows that the output from the 7555 ranges from almost exactly 0V to 9V, when you are using a 9VDC power supply. This voltage will be pulled down somewhat when you add a load such as a speaker to the output pin of the chip. The oscilloscope trace shown in Figure 1-8 was measured without a load.

The **period** of the output is a measurement of **time**, from the start of one pulse to the start of the next. In other words, the period is the duration of the pulse, plus the duration of the gap between it and the start of the next pulse. In this example, the period is measured in milliseconds (ms). A millisecond is $\frac{1}{1000}$ of a second.

When the signal from the 7555 passes through the speaker, the **cone** of the speaker (also known as its **diaphragm**) vibrates and converts each pulse into a pressure wave in the air. When a membrane in your ear responds to these pulses, your brain interprets that stimulus as sound.

The **frequency** of a signal is the number of pulses per second. This is expressed in **hertz**, named after the electrical pioneer Gustav Ludwig Hertz, and is abbreviated as *Hz* (the *H* is capitalized because it refers to his name.) A frequency of 100 pulses per second is written as 100Hz, while 1,000 pulses per second would be 1 **kilohertz**, written as 1kHz, and 1,000,000 pulses per second are 1 **megahertz**, written as 1MHz. The capital *M* means "mega." (A lowercase *m* in the metric system means "milli," so you must avoid getting them mixed up.)

Human hearing can resolve sounds ranging from around 20Hz to a maximum of 20kHz, although elderly people may have difficulty hearing sounds higher than 10kHz, and people who have damaged their hearing as a result of environmental noise (or attending rock concerts) may have a limit as low as 5kHz.

A sound wave, consisting of alternating regions with high pressure and low pressure, travels through the air at approximately 1,125 feet per second at sea level. (At higher altitudes, sound travels more slowly.) The distance from the start of a high-pressure region to the start of the next high-pressure region is called the wavelength. (Remember, the period is a measurement of time, while the **wavelength** is a measurement of distance.)

If **f** is the frequency of a sound, measured in Hz, and **p** is the period in seconds, and **w** is the wavelength in feet, and **s** is the speed in feet per second, these values are related to each other by two simple formulas:

$$f = 1 \; / \; p$$
$$s = w \; * \; f$$

(Throughout this book, we will use the / symbol as a division sign and the * symbol as a multiplication sign.)

You can see from the scale at the bottom of Figure 1-8 that the period of this soundwave is about 3.5 milliseconds, which is 0.0035 seconds. Using the first formula, **1 / 0.0035 = 286Hz** (approximately). That's the frequency of this sound. If you rewrite the second formula as

$$w = s \; / \; f$$

you can see that the wavelength of the sound is **1,125 / 286**, which is about 4 feet.

THE CARRIER WAVE

In theory, you could disconnect the speaker from your circuit and substitute a piece of wire, which would function as a ***transmitting antenna***. This would radiate a small amount of power, which you would be able to pick up by using another circuit with a ***receiving antenna***.

In practice, higher frequencies can transmit more power over longer distances. In any case, if all radio stations in the world transmitted audio frequencies, we would have no way to separate them and listen to just one source at a time.

The answer to both of these problems is to add the audio frequency to a much higher frequency, known as a ***carrier wave***. Then each radio station in its area can use a different carrier frequency so you can filter out the ones that you do not want to hear. If you look at the AM tuning dial on a radio, you'll see numbers ranging from around 540kHz to 1,600kHz, which are carrier frequencies.

To add your audio soundwave to a carrier wave, all you have to do in this experiment is connect a second 7555 timer running at a higher frequency. Strange fact: A 7555 can run at up to 2MHz (which is the same as 2,000kHz), so it is quite capable of transmitting a radio signal in the AM waveband, even though it is never normally used for that purpose.

First, remove the speaker from the circuit that you just built. You can also remove resistor R3 and capacitor C4. Then, add the extra 7555 timer, as shown in Figures **1-9** and **1-10**. Components that you placed on the

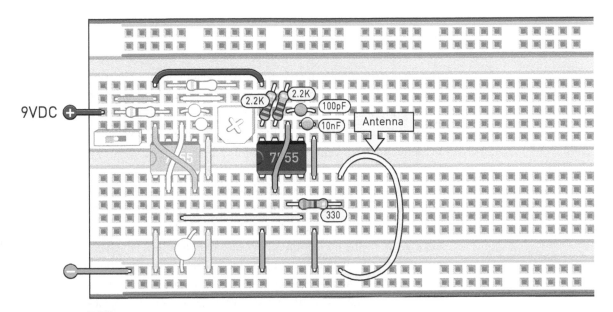

1-9 *Adding a second timer that runs at a radio frequency.*

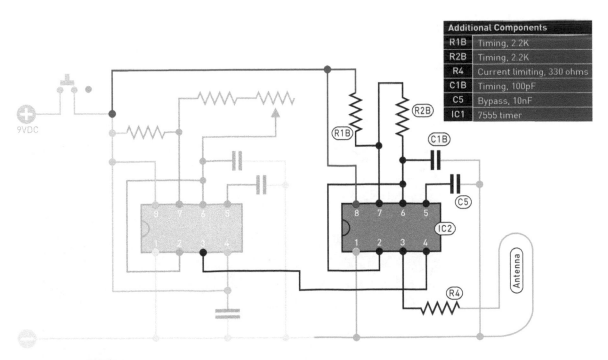

Additional Components	
R1B	Timing, 2.2K
R2B	Timing, 2.2K
R4	Current limiting, 330 ohms
C1B	Timing, 100pF
C5	Bypass, 10nF
IC1	7555 timer

1-10 *The schematic version of the breadboard circuit in Figure 1-9.*

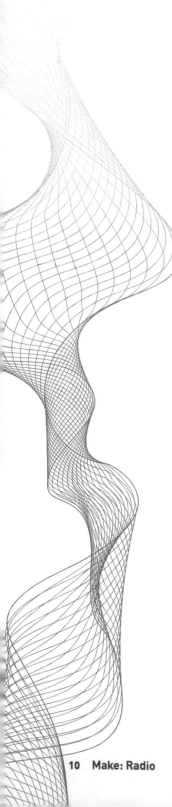

breadboard previously are still there and will still be necessary, although they are grayed out right now.

The new component values in this circuit will generate a carrier frequency around 800kHz. The circle of yellow wire in Figure 1-9, connected through a 330-ohm resistor with the output from IC2, is a loop about 2" in diameter, which will function as your transmitting antenna. The 330-ohm resistor prevents IC2 from being overloaded.

Here's how the circuit works: Pin 4 of the 7555 timer is the reset pin, which puts the timer on hold when the pin voltage is near zero (negative ground) but allows the timer to run when you apply a voltage near the power supply. Notice the horizontal yellow wire in Figure 1-9, which connects the output pin of IC1 with the reset pin of IC2. This means that the output of IC1 switches IC2 on and off. Remember, IC1 is running at an audio frequency. IC2 doesn't mind being switched on and off quickly, even at several kilohertz.

Notice that C1B, which you can see in Figure 1-10, is 100pF. (That's picofarads, not nanofarads. Be careful not to get your *p*'s and *n*'s mixed up.)

When you apply power to the circuit, you will not be able to hear anything, because the speaker has been removed. The loop antenna is broadcasting a signal, and in a moment. I will show you how to pick it up and make it audible.

While your new circuit is running, IC2 generates a carrier wave of around 800kHz. This is switched on and off by IC1. In Figure **1-11**, an oscilloscope

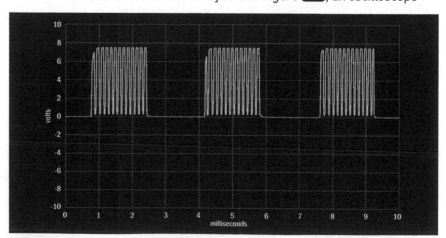

1-11 *Output from the second timer. The oscilloscope has downsampled the high frequencies to make the oscillations individually visible.*

was attached to the output from IC2 so that you can actually see how this happens. (In reality, there would be as many as 1,000 oscillations inside each burst, rather than the dozen or so shown in Figure 1-11. The oscilloscope downsamples them so that you can see what is going on.)

What if you don't have an oscilloscope? How will you know that IC2 is actually doing anything?

If you have a portable AM radio, try tuning it slowly through its range while holding it near the yellow loop antenna on the breadboard. At some point, you should hear the radio clearly receiving the audio frequency that has been superimposed on the carrier wave.

Congratulations! You just demonstrated a radio transmitter. But wouldn't it be more interesting to build a real receiver of your own?

This is easily done.

LOCAL RECEPTION

Two main types of transmission are used by radio stations today: **frequency modulated** (abbreviated as *FM*) and **amplitude modulated** (abbreviated as *AM*). The earliest radio transmissions were AM, meaning that the loudness of the sound coming out of the radio was proportional with the **amplitude**, or voltage, of the signal. I'll get to FM radio later in the book. Right now, you're going to build AMR1, your first AM radio receiver. Long ago, this type of radio was referred to as a crystal set, because it used a semiconducting crystal as a diode before diodes became available as components.

To build it, you only need five components: a suitable coil, a suitable capacitor, a suitable diode, a suitable resistor, and a suitable earphone. First, you'll assemble the components and use them to tune in to your 7555 timers so you can make sure that everything functions properly. Then I'll explain how it all works.

The coil in AMR1 will be a length of wire wrapped neatly around a ferrite rod. (The rod is illustrated in Figure **1-12**.) Ferrite just happens to be a substance that intensifies the magnetic field created by a coil of wire.

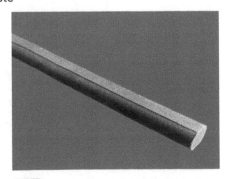
1-12 *A ferrite rod at least ⅜" in diameter.*

For the coils in this book, we'll be using 26-gauge hookup

wire—not the usual 22-gauge—because we want to pack the turns of wire closely together. The color of the insulation doesn't matter, but make sure that the copper inside the insulation is solid, not stranded.

You may wonder why I'm not using **magnet wire**, which is copper wire with the thinnest possible insulation. This can maximize the power of a coil, as it can be very tightly packed, but I decided to avoid the extra expense of magnet wire. You would also have more difficulty removing its insulation, as it must be scraped off with a knife or with fine sandpaper.

1-13 *Step 1 in creating a coil is making a cardboard sleeve to wrap around a ferrite rod.*

You will need to wind your coil over a sleeve of thin card stock that can slide up and down the rod so that you can adjust the effect of the coil. Wrapping the card around the ferrite rod can be done in four steps:

First, cut a 3" x 1.5" piece off a file card. Add a piece of clear adhesive tape along one long edge. In Figure **1-13**, we used black electrical tape, as it shows more clearly in a photograph, but clear tape will hold the card more firmly.

1-14 *Creating a curl in the cardboard.*

1-15 *Wrapping the card around the rod.*

Then, wrap the card tightly around a thin rod such as the shaft of a screwdriver, as in Figure **1-14**, but don't allow the tape to stick. You're just putting a curl into the card.

Now, place the ferrite rod inside the curve of the card, as in Figure **1-15**. Tighten the card around the rod and then allow it to loosen just a fraction so that you'll be able to slide it freely. Tape the card in place as in Figure **1-16**.

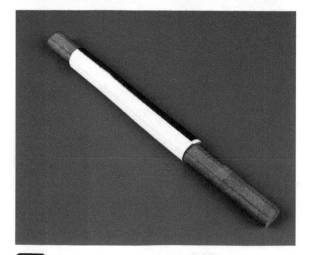

1-16 *The card taped around the rod, with a little gap to allow it to slide.*

Use another piece of tape to secure your wire to one end of the file card, allowing 6" of wire to hang free. Now, wrap 40 turns of wire around the card. This will be easiest if you are able to clamp one end of the rod in a vise, although not too tightly, bearing in mind that the rod is brittle and can break. A vise with plastic jaws is best.

You may find it easiest to apply 10 turns of wire loosely, then push them together and rotate them to tighten. Then add another 10 turns, and another, and another. Finally, use a second square of tape to secure the coil to the card, with 6" of wire hanging free at the end. The result should look like Figure **1-17** if you do it carefully.

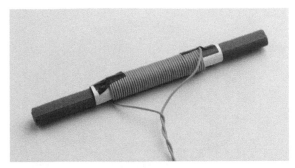

1-17 *The completed coil.*

Next, you will need a *variable capacitor*, also known as a *tuning capacitor*. If you ever open up a big, old AM radio, you may find a variable capacitor consisting of rigid aluminum plates, such as the example in Figure **1-18**. One set of plates is fixed, while the other set rotates when you turn a tuning knob. The amount of overlap between the sets of plates can vary from near 0% to near 100%.

1-18 *A classic-style variable capacitor.*

Old-style capacitors of this sort are still available secondhand from sources such as eBay. They are sometimes known as *air-gap capacitors* because the plates are separated by air. They can be expensive, and the seller may not know what the capacitance range is. The one I want you to use, as shown in Figure **1-19**, is cheaper and much more compact.

It contains very thin flexible plates separated by thin wafers of plastic and sealed inside a plastic case. This particular component is available from multiple sources and actually contains two capacitors on the same shaft, with maximum values of 140pF and 60pF. It has a part number of 223P; don't order a 223F by mistake, as it has

1-19 *The smaller and less expensive variable capacitor recommended for projects in this book.*

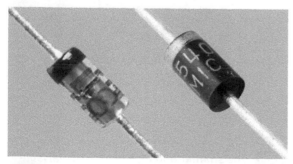

lower capacitance. We will connect the internal capacitors in parallel to create a maximum capacitance of 200pF.

If your capacitor was supplied with a wheel that fits on the shaft, this will make it easier to use. You may also find the wheel sold as a separate accessory. Attach the wheel by pressing it down and then securing it with a screw, as shown in Figure **1-20**. (The rest of my photographs don't show the wheel, because it prevents you from seeing the connections of the capacitor.)

1-20 *A wheel allows you to tune the capacitor more easily.*

On the back side of your capacitor, you will find four little nuts and two screws. Each screw is connected with a tiny semicircular trimmer capacitor, visible through the transparent shell of the component. You can see them in Figure **1-21**. Each trimmer can be used to adjust the tuning range. Use a small, flat-bladed screwdriver to set them at their minimum overlap, as in the photograph.

1-21 *Trimmers on the underside of the capacitor.*

Now, you will need a **diode**, which allows current to pass in one direction while blocking it in the other direction. **Silicon diodes** are commonly used in electronics, but they require a minimum of 0.7V, which AMR1 cannot provide, as it won't have its own power supply. You need a **Schottky diode** or a **germanium diode**, either of which will work with only 0.3V. Two examples are shown in Figure **1-22**. (Bear in mind, each of them is only about ¼" long.) While diodes vary a lot in appearance, one end is always marked with a stripe, which may be light or dark but always means the same thing: This is the end from which conventional current (positive to negative) **flows out**. In other words, the diode conducts when the marked end of the diode is "more negative" than the unmarked end.

1-22 *Two types of diodes, each about ¼" long.*

The low voltage in AMR1 imposes another requirement: You need an earphone that has a **high impedance**. The earphones that people use with a music player or a phone are **low impedance**, and will not work here. (The term *impedance* refers to the varying resistance that a component can present to a varying current.) A full discussion of earphones is in the shopping list in Appendix A. You may choose to use a **passive piezo transducer** (a miniature loudspeaker) if you have difficulty finding a high-impedance earphone.

The schematic for AMR1 is shown in Figure **1-23**. I suggest not mounting the components on your breadboard, because we want to demonstrate that your radio receiver works without any power from the board. We also want to make it easy to use the tuning capacitor, which has flat contacts that won't fit into a breadboard. The answer to these problems is a European-style **terminal block**, such as the one shown in Figure **1-24**.

This one happens to have three pairs of terminals, which are each equipped with a pair of screws (one per terminal) that you can tighten with a miniature screwdriver. In the photograph, just for demonstration purposes, the white wire is connected through the block to the black wire, while the yellow wire is connected through the block to the red wire. The block keeps each pair of wires isolated from the other pair.

For our projects, eventually, you will need twelve pairs of connections. Fortunately, you can buy a block of this size, which is designed to be chopped into pieces with a utility knife, as in Figure **1-25**. You will need to cut the block into three pieces: one with three pairs of terminals, one with four pairs of terminals, and one with five pairs of terminals, as

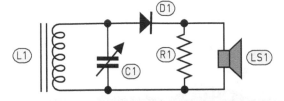

Components	
L1	Coil, 40 turns on ferrite rod
C1	Tuning capacitor, 200pF
D1	BAT48 Schottky diode or similar
R1	10K resistor
LS1	High-impedance earphone

1-23 *The schematic for ultrasimple receiver AMR1.*

1-24 *For demonstration purposes, this terminal block is connecting the white wire with the black wire and the yellow wire with the red wire.*

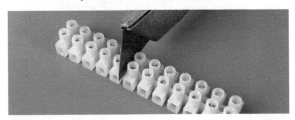

1-25 *You can use a utility knife to cut a terminal block into segments.*

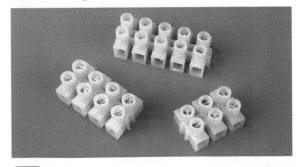

1-26 *You will need these segments of a terminal block.*

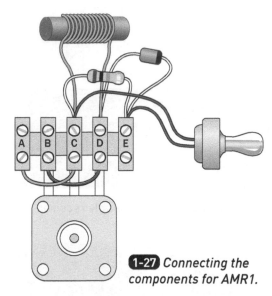

1-27 *Connecting the components for AMR1.*

1-28 *Getting ready to connect the variable capacitor.*

1-29 *Squash the contacts so that wire can be wrapped easily around them.*

shown in Figure **1-26**. You will use the five-terminal segment in this experiment, but please keep the other pieces for use in the next experiments.

Terminal blocks come in many styles and sizes, but the one you need has terminals spaced at $\frac{5}{16}$" to match the spacing of contacts sticking out of the tuning capacitor. (The spacing of terminals in a block may be referred to as their **pitch** in a datasheet.)

Figure **1-27** shows how you will connect your coil, tuning capacitor, 10K resistor, diode, and earphone through a block with five pairs of terminals. If you compare this diagram with the schematic in Figure 1-23, you'll see that the wires have been moved around. However, the connections are the same. Your goal is to push two or three wires into each hole in the block and tighten the screws to hold everything together. That sounds easy enough, but unfortunately, wires have a habit of coming loose, especially if you transport your AMR1 from one place to another.

To secure the wires, you can twist them together before you insert them in the block. I'll take you through this process step by step.

First, we will deal with the variable capacitor. A couple of wires must be added to its terminals, for reasons I'll explain in a moment. (The parts are shown in Figure **1-28**.) To make a really secure connection, squash the contacts of the capacitor with pliers before wrapping wires around them.

Figure **1-29** shows the contacts after they have been squashed. Figure **1-30** shows how the end of the hookup wire has been twisted around one of the contacts. Figure **1-31** shows the remaining connections. The purpose of the red wire is to join the two capacitors that coexist inside the component. The purpose of the blue wire will become clear.

1-30 *Wrapping a wire around one of the contacts.*

1-31 *Ready for insertion into the terminal block.*

1-32 *The complete assembly on one side of the terminal block.*

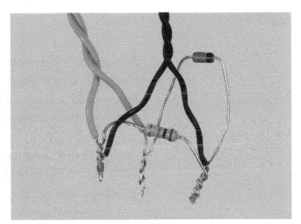

1-33 *Components prepared for installation on the other side of the terminal block.*

Figure 1-32 shows the capacitor and its wires installed in the terminal block. Notice in Figure 1-27, the blue wire connects terminals A and C, while the red wire connects terminals B and D. Figure 1-32 shows the actual assembly.

Now for the connections on the other side of the block. You can twist the wires together, as shown in Figure 1-33, so that the connections are the same as in Figure 1-27.

The black wires go to the earphone, the green wires go to the coil that you created, and the resistor is 10K (with brown, black, and orange stripes). The diode must have its marked end facing toward the right, but the earphone, coil, and resistor can be connected either way around. Be careful not to short the leads going into the resistor against the wires underneath it.

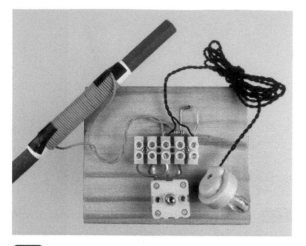

1-34 *The complete AMR1 circuit.*

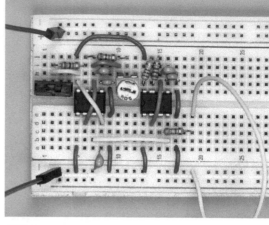

1-35 *The breadboarded transmitter circuit.*

The complete assembly is shown in Figure **1-34**. I screwed the terminal block to a piece of wood because this makes everything easier to handle. A couple of #2 screws will fit. If your local hardware store doesn't have screws as thin as this, you can use double-sided adhesive, glue, or some duct tape.

The upper half of each terminal A and B is not being used for anything—but they'll have a purpose after you test the circuit. The bottom half of terminal E is empty, and it will stay that way because we just wanted to use the top half as a convenient way to connect three wires.

Go back to your transmitter circuit and make sure that it is powered up. It should look like Figure **1-35**, in which the loop of yellow wire at the bottom is the transmitting antenna.

Insert the earphone in your ear. Don't try to apply power to your AMR1 assembly, because it doesn't need any. Simply take hold of the ferrite rod with the coil around it and insert it into the loop of yellow wire. Turn the shaft of the variable capacitor, and near the middle of its scale, you should hear the audio generated by IC1. Rotate the trimmer potentiometer, P1, on your breadboard, and the complete range of audio frequencies should be audible.

Pull the ferrite rod out of the loop antenna, and the sound diminishes in volume. The sound will also diminish if you slide the coil away from the center of the rod.

This simple demo tells you something quite important: *Power does travel through thin air.*

The output from IC2 on your breadboarded circuit is only around 30 milliamps (mA) at 7.5V, but somehow it travels out of the antenna (the loop of wire), through the air, and into the coil on the ferrite rod. From there, it passes through a diode and into the earphone, where it vibrates a diaphragm with enough energy for you to hear it.

You can experiment by substituting a 10pF capacitor instead of a 100pF capacitor for C1B in your transmitter circuit in Figure 1-10. This will increase the carrier frequency, and you will find that the variable capacitor on your receiver is now more sensitive toward the end of its range. (A portable radio will show the same sensitivity.) This is an important discovery: *The variable capacitor tunes the receiving circuit to match the carrier frequency of the transmitting circuit.*

The next step will be to modify AMR1 so that it will pick up broadcasts from radio transmitters that may be many miles away. Before you move on to that however, I need to explain a couple of features in the circuit.

HOW IT WORKS

First, in the schematic in Figure 1-23, there is the diode labeled D1. What is it for?

Flip back to Figure 1-11, which shows the output from IC2. This is the signal that goes into the loop of wire, which is your transmitting antenna. The oscilloscope trace shows that the output of IC2 varied between 0V and approximately 7.5VDC.

Now, take a look at Figure **1-36**. This trace was measured between the two wires at the ends of your receiving coil in AMR1 (screws C and D in the terminal block shown in Figure 1-27). Somehow, the coil picks up the transmission and turns it into a voltage that varies between −0.35V and +0.35V. Why is the voltage now shown as varying between a positive value and a negative value?

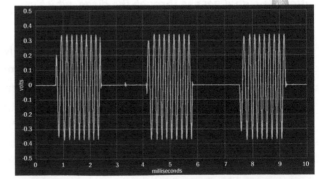

1-36 *Voltage across the receiving coil in AMR1.*

Because voltage is always relative, and only variations in current get transferred from the transmitter to the receiver. Therefore, the signal from the coil in the ferrite rod fluctuates around 0V.

Now, what would you hear if the diode wasn't in the circuit? Nothing! You can test this yourself by shorting out the diode with a piece of wire between screws D and E in Figure 1-27.

The diaphragm in the earphone cannot vibrate at the carrier frequency. The oscillations are much too fast—and even if the earphone could reproduce them, they are too fast for your ear to resolve them as sound. Moreover, the oscillations are equally positive and negative, so they add up to an average of zero.

When you pass the output from the coil through a diode, it blocks the negative half of each oscillation, and only the positive part flows through. This process of allowing current to flow in one direction, but not the other direction, is known as **rectifying** the signal.

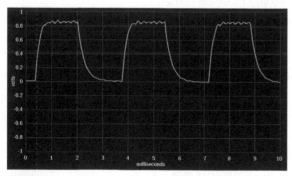

1-37 *Voltage across the earphone in AMR1, after the signal passes through the diode, which blocks negative fluctuations.*

You can think of all the tiny, rapid, positive pulses of the carrier frequency pushing the diaphragm of the earphone in one direction. Then there's a pause (lasting a fraction of a second), and the diaphragm relaxes. This happens between 200 and 3,000 times per second, at the audio frequency set by IC1. You can see this in Figure **1-37**, which is an oscilloscope trace taken from the wires to the earphone (screws C and E in Figure 1-27). *The earphone rides the peaks of each burst of carrier frequency.*

Next question: How does the tuning capacitor match the frequency of the receiver to the frequency of the carrier wave? This involves the concept of electrical **resonance**. Basically, the combination of the coil and the capacitor will resonate at a certain frequency, and when that frequency matches the frequency of the transmitter, they respond. A simple analogy may make this clear.

1-38 *An experiment to demonstrate resonance.*

FUNDAMENTALS OF RESONANCE

You don't need electricity to see an example of resonance. A couple of weights and some pieces of string will provide a demo. The weights can be made of anything; I suggest using plastic water bottles that are at least half full. Stretch a piece of string horizontally between the backs of two chairs, as shown in Figure **1-38**, and suspend bottles by vertical strings. Nudge one of the bottles to start it swinging, and if the other bottle hangs from a different length of string, it will barely respond. Now, adjust the vertical lengths of string so that they are equal, and the motion of the first bottle will cause the second bottle to swing in sync. The second bottle is resonating with the first.

In electronics, we can use a coil and a capacitor to create a resonance circuit that will oscillate at a particular frequency, like a bottle hanging from a particular length of string. In the receiver circuit that you just built, the coil and capacitor were chosen to resonate with the frequency of IC2 on the breadboard. If you changed capacitor C1B by substituting one with a value of 10pF, the carrier frequency would change, the receiver wouldn't resonate with it anymore, and you wouldn't hear anything.

But how, exactly, does an electronic circuit resonate?

COILS AND CAPACITORS

When current passes through a coil, it creates a magnetic field in the center of the coil. The magnetic field seems to occur instantly, but actually, there is a tiny delay determined by the size of the coil and the number of turns of wire. During that delay, current will not pass through the coil, because electrical energy is being used to build the magnetic field.

A capacitor behaves in an opposite way. An initial pulse of current passes through the capacitor before the capacitor builds a stable charge on its plates. The initial pulse is sometimes referred to as *displacement current*.

SUMMING UP:
- A coil tends to block current initially, and then passes it.
- A capacitor passes current initially, and then blocks it.

What if you put a coil in parallel with a capacitor, as in Figure **1-39**? This is exactly how the variable capacitor and the coil around the ferrite rod are wired in AMR1. This

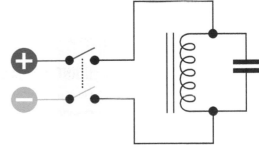

1-39 *An electronic resonance circuit.*

basic circuit is sometimes known as a **resonant circuit**, a **tank circuit**, a **tuned circuit**, or an **LC circuit**. (In the *LC* acronym, *C* refers to capacitance, and *L* refers to inductance because a physicist named Heinrich Lenz pioneered the understanding of inductance during the 1800s.)

In the setup shown in Figure 1-39, if you press and release the switch very, very briefly, the capacitor charges while the coil is busy generating its magnetic field. Then the coil allows current to flow, so the capacitor discharges—but when the capacitor has no more power to supply, the magnetic field in the coil collapses, which releases electric current, and this recharges the capacitor with opposite polarity. In this way, the circuit oscillates, with electric current surging like water sloshing from side to side in a tank (which is where the term "tank circuit" comes from).

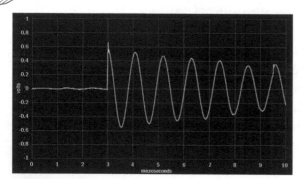

Eventually, after you release the switch, the oscillations will fade away because the circuit contains some resistance. Figure **1-40** actually shows this happening when the oscilloscope was connected across the coil in the AMR1 circuit and a momentary blip of voltage was applied.

1-40 *Oscillations in a resonant circuit.*

You can compare this with the model I described earlier, featuring two weights hanging on equal lengths of string. When you nudge one weight, its motion feeds through to the other. But if you don't provide a continuous input, the oscillations of the weights will gradually diminish because of air resistance and other factors.

But suppose you could hit the switch in Figure 1-39 repeatedly, with superhuman speed, in sync with the oscillation of the circuit. You could keep nudging the oscillation to keep it going. On the other hand, if you hit the switch out of sync, you would interfere with the oscillations, and they would stop.

DOING THE MATH

Do you wonder how the values for a coil and a capacitor can be chosen to resonate at a particular frequency? There is a formula to calculate this. In the formula below, **L** is used for **inductance**, which is measured in **henries**, named after Joseph Henry, yet another electrical pioneer. **C** represents capacitance in farads, and **f** represents the **resonance frequency** in hertz.

The term **sqrt** means "take the square root of the function that follows in parentheses." The term **pi** is the same value as you would use for calculating the circumference or area of a circle: approximately 3.142.

```
f = 1 / (2 * pi * sqrt(L * C))
```

Remember, throughout this book, I'll use **∗** as a multiplication sign to avoid confusing it with the variable **x**. The **/** sign indicates division. In any formula containing parentheses, you do the calculation inside the parentheses first. If there is a pair of parentheses inside another pair, you do the calculation inside the most deeply nested pair and work your way out.

In radio, where high frequencies are used, the formula is more useful if you multiply both sides by 1,000,000. Now, **f** is the frequency in MHz. On the right-hand side, because the values are divided into 1, **L** is now the inductance in **μH** (microhenries) and **C** is the capacitance in **μF** (microfarads).

But how do you know how many turns of wire to use, and what their diameter should be, to create the inductance that you need to provide the frequency that you want?

There's a formula for this, too, but it's only approximate. In fact, it is known as **Wheeler's approximation** and is illustrated in Figure **1-41** (reproduced from **Make:Electronics**). This figure also shows some examples of inductance using a variety of turns of wire on a spool of a certain size. Bear in mind that these numbers only apply when the coil has an open core and the coil is wound around a nonmagnetic frame. As soon as you insert a ferrite rod, the inductance increases.

LONG-DISTANCE RECEPTION

Now that you have verified that your little AMR1 circuit can pick up a signal generated by the timers in the test circuit, it's time to tune in to some real radio signals. Not many people still

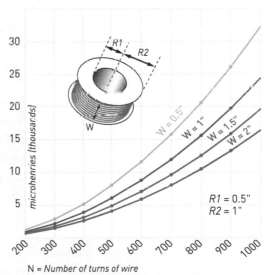

N = Number of turns of wire

From Wheeler's approximation formula:

$$\text{Microhenries} = \frac{0.8 * A * A * N * N}{(6 * A) + (9 * W) + (10 * D)}$$

Where A (average winding radius) $= \dfrac{R1 + R2}{2}$

and D (radius difference) $= R2 - R1$

1-41 *Some examples of inductance in four coils of wire, and an approximate formula. All lengths are in inches.*

listen to AM radio, but stations are still out there and are still broadcasting. The only problem is, AMR1 isn't sensitive enough to hear them in its current form. We have to fix that.

The word **antenna** is confusing because an antenna can either transmit radio waves or receive them. Previously, in your test circuit, you used a transmitting antenna consisting of a single loop of wire. Now, you need to attach a much longer piece of wire to your receiver circuit as a receiving antenna.

I have to warn you in advance that this doesn't always work with AMR1, because we have not yet added a power supply to amplify the output. You need to be reasonably close to an AM radio station, and you'll have a much better chance of picking up signals at night, as AM travels farther after sunset. When it does work, though, it can seem miraculous.

(If you are wondering why the Sun interferes with radio reception, this is because it overwhelms our transmissions with its own electromagnetic radiation.)

Your receiving antenna must be as long as possible. I think 30 feet should work, and heavy wire may achieve better results than thinner wire. If you happen to have the kind of 14-gauge wire that is sold for use as 120V wiring in your house, that will be ideal, and it is available from any hardware store. It can be stranded or solid; either is good. If you only have 22-gauge hookup wire, try that. If you have multiple short pieces of wire, you can strip the ends and join them in series.

The antenna should not touch any objects that will tend to connect it with the ground. (I'm talking about the actual, physical ground of planet Earth, not the negative ground on your circuit board.)

Ideally, you should conduct this experiment outside somewhere—which should be easy, as AMR1 requires no power. (If you live in an urban area, go to a park.) Tie a weight to a piece of string, and tie the other end of the string to the wire that you will use as your antenna. Throw the weight up into a tree, and orient the antenna so that the wire doesn't touch the tree. (That's what the string is for.) You will connect the lower end of your antenna to AMR1, as I will describe in a moment.

CAUTION: Never string an antenna outside if there is any risk at all of thunderstorms. Benjamin Franklin survived when he flew a kite in the middle of a storm. He was lucky. You might not be. Being struck by lightning is extremely hazardous to your health.

If you want to perform your experiment indoors, you can hang the antenna wire from pieces of string attached to light fixtures or window shades.

You will also need a ground wire—and here, again, I am referring to the actual ground. Ideally, if you're outside, you would run a wire to a metal stake that you hammer into soft, moist earth. Alternatively, look for any large metal object that makes contact with the ground. A chain-link fence would be ideal (although not if it is the type where the metal is coated in plastic).

If you are indoors, you can ground your receiver by attaching a wire to any metal object that has a connection, ultimately, with the ground outside. You could use a steel or copper water pipe, for example. (Plastic water pipes will not work.) Alternatively, you can attach your ground wire to the steel cabinet of an appliance such as a washing machine, which is grounded through the electrical system in your home. Similarly, a high-end audio receiver may be in a steel case and may have a ground terminal at the back.

To make a connection with an appliance, you may be able to find a screw securing the steel case to the internal chassis. You can loosen this screw and attach a ground wire under it. Or you can drill a small hole and insert a screw of your own. (If you are a young person reading this book, you may want to check with adults in the house before you start drilling holes in appliances.)

Maybe you are thinking that all the electrical outlets in your home are grounded. Indeed, this is true: That is what the little round hole in an outlet is for, although in countries outside of the US, it may be a little rectangular hole or an exposed metal piece. The problem is, there's also a live socket in each power outlet, very near to the ground connection, and it's potentially lethal. I don't want you to risk a high voltage traveling up the wire to your earphone and frying your brain. Accidents can happen, and therefore:

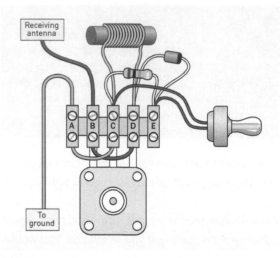

1-42 *Adding antenna and ground connections to AMR1.*

Now, supposing you have your antenna and your ground wire, where do you attach them to AMR1? Take a look at Figure **1-42**. You can see why I provided those empty connectors labeled A and B. Also, you see why a red wire was used previously: It's the source of power for this circuit, derived from the radio signal that you receive. And the blue wire connects with the ground.

The schematic in Figure **1-43** has been modified with an antenna symbol (looking like an umbrella that has blown inside out) and a ground symbol (at the bottom of the circuit). These are not the only symbols that people use for antenna and ground, but they are relatively common in the United States, and are recognized elsewhere. Note that in some countries, especially the UK, the ground connection is referred to as the **earth**.

Compare Figure 1-43 with Figure 1-39. Aha! You see now that if the radio signal has a frequency that matches the LC circuit, the LC circuit will resonate with it.

TESTING, TESTING . . .

If you are trying to pick up radio stations with AMR1 inside your home, the next step is to make sure that it will not be affected by unwanted electrical noise. Switch off the test circuit that you built around the 7555 timers, and consider that some appliances may create interference. A lamp dimmer, for instance, can introduce a frequency into your house wiring that AMR1 will pick up as a buzzing sound. Fluorescent lights also create interference. Any

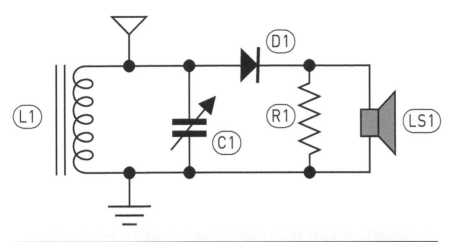

Components

L1	Coil, 40 turns 26-gauge wire, on ferrite rod, 3/8" x 6"
VC1	Variable capacitor, 200pF
D1	Schottky or germanium diode
R1	10K resistor
LS1	High-impedance earphone or passive piezo speaker

1-43 *Antenna and ground symbols added to the schematic for AMR1.*

buzzing noise in the earphone attached to your radio probably comes from a domestic appliance. To get rid of the buzzing, you will have to go around switching things off until you do not hear the buzzing anymore.

Now, with the earphone inserted into your ear, slowly turn the tuning capacitor through its full range, and remember that your chances of hearing something will be much better after sunset.

Because a lot of AM stations broadcast spoken-word programs, you are likely to hear a human voice. Someone, probably many miles from your home, is speaking into a microphone in a studio. Their voice is amplified at the radio station and merged with a carrier wave, and the signal goes up a tall steel transmission tower. From there, a slightly mysterious field effect carries the power through the air, where it induces just enough current in your antenna to vibrate the diaphragm in your earphone, without any extra electricity required.

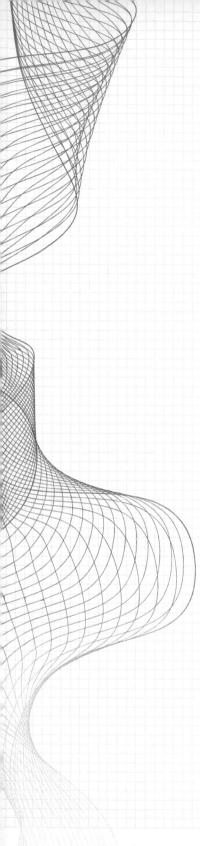

A WORD ABOUT TESLA

It's easy to imagine how Nikola Tesla made an intuitive leap and wanted to crank up a radio transmission so that it would broadcast serious electrical power. His most ambitious installation, in Colorado Springs, used a coil 100 feet in diameter, which created a potential of up to 12 million volts. It broadcast sufficient power to light a fluorescent tube equipped with an antenna 25 miles away. It also spooked horses by inducing electric shocks in the iron shoes on their hooves.

This was a time when some aspects of electricity and physics were not fully understood, and there were competing theories to explain them. Unfortunately, Tesla pinned his hopes on theories that turned out to be wrong. He believed empty space wasn't really empty but was filled with an invisible, mysterious substance known as *ether*. He hoped that if he tapped into the ether at the right frequency, infinite power was available. This turned out not to be the case.

If he had devoted less time and effort to his dream of infinite electrical power, probably he could have won the race to transmit a radio message across the Atlantic. Marconi ended up with that achievement, but Tesla had a more powerful transmitter, a larger antenna, and a more sophisticated receiver. He just didn't think that radio messages were as interesting as free electricity.

WHY DIDN'T I USE THE BOTTLE VERSION?

I hesitate to annoy you by mentioning *Make:Electronics* again, but since you may own a copy, I have to explain why I didn't build AMR1 in the same way as the unpowered radio project in that book. In that project, a coil was wound around a plastic bottle, and no ferrite rod was required. Moreover, there was no capacitor. Since that seems simpler, why didn't I use it here?

The reason is that the ferrite rod in AMR1, coupled with the tuning capacitor, achieves better results. I have tested AMR1 against the bottle radio, and the sound from AMR1 is louder and clearer.

But how does a bottle radio work with no capacitor? The answer is that a long antenna generates its own capacitance between itself and the ground. It's not adjustable, though, so the coil on the bottle radio includes *taps* consisting of little loops in the wire. This allows you to select as much of the coil as you want, enabling a primitive system for tuning the radio.

An example of the bottle radio is shown in Figure **1-44**. I include it here because if I didn't mention it, you might wonder why. The corresponding schematic is in Figure **1-45**. The taps are shown as little dots on the coil.

TWEAKING AMR1

Because AMR1 is operating at the limits of feasibility, you may need to tweak it a little to make it perform well.

Your antenna will add capacitance to the circuit, depending how long it is and how it is oriented. This will shift the sensitivity of the receiver toward lower frequencies. You can compensate by reducing the coil inductance; just slide the coil toward one end of the ferrite rod (or even slightly beyond the end).

With a coil consisting of 40 turns of 26-gauge wire on a 6" ferrite rod, and a 30-foot antenna, I found that the receiver was sensitive in the following ranges:

Frequency range (kHz)	Coil position
650–970	Middle of rod
744–1100	End of the rod
1200–1650	⅓ over the edge

AM broadcasts in the Americas range in frequency from slightly above 530kHz to slightly below 1,700kHz.

Incidentally, some early radio sets were tuned entirely by moving a ferrite core inside a coil.

You can adjust the number of turns in the coil,

1-44 A bottle radio is simpler than AMR1 but doesn't work as well.

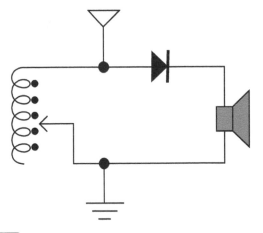

1-45 The schematic for the bottle radio in Figure 1-44.

bearing in mind that when you add turns, the higher inductance reduces the resonant frequency.

You can also try different diodes. At a location that is relatively far from AM transmitters, a BAT48 Schottky diode and an AA119 germanium diode perform similarly. Nothing was audible when I used a 1N4148 silicon diode, which was not a surprise, because of its higher threshold voltage. With another type of Schottky diode, the 1N5817, nothing was audible—so evidently not all Schottky diodes function interchangeably.

If you are in an urban area, you may hear more than one radio station, with their signals merging, because AMR1 isn't very selective. If you live in a rural area, you may have difficulty picking up any signals at all, although you should try sliding the coil up and down the ferrite rod while you also adjust the tuning capacitor, and remember to do your listening after sunset.

The surprising aspect is that AMR1 will deliver any sound at all despite having no power supply and an earphone that is entirely driven by radio waves. Since your little receiver is taking its power from a radio station, does that mean it actually adds to the load on the transmitter at the station? The answer is yes—but to such a small extent, no one would notice the difference.

ADDITIONAL OPTIONS

You may try using an output transformer to match the output from your LC circuit to your earphone. The Eagle P631M (formerly LT-44) is an example. You can find sources online for other unpowered radio circuits, including these:

- web.archive.org/web/20211217082237/makearadio.com/
- crystal-radio.eu/index.html
- hobbytech.com/crystalradio/crystalradio.htm
- www.usefulcomponents.com/main_contents/projects/choccy_block
  _crystal_radio/choccy_block_crystal_radio.html

STATUS UPDATE
What's the story so far, and what comes next?

This chapter began by demonstrating the concepts of frequency and wavelength, using a simple oscillator circuit to generate audio frequencies and another oscillator to create a carrier wave. You learned about ferrite rods, handmade coils, variable capacitors, diodes, and how to combine these components to create a basic unpowered receiver named AMR1. It demonstrated the concept of resonance in electronics.

This is as far as we can go with AMR1. The next step is to build AMR2, which will use battery power and transistors to achieve three goals: loudspeaker output, sensitivity to distant signals, and better selectivity to choose among them.

experiment

2

A REAL RADIO

In Experiment 1, the little receiver that we named AMR1 was only suitable for demonstration purposes. It detected the transmitter on your tabletop, and It could pick up local radio stations, but it couldn't tune in to transmissions from far away.

To make those distant signals audible, you will need a more ambitious circuit that adds some **amplification**. The easiest way to achieve this is with some everyday **bipolar transistors**. In case you are not completely familiar with transistors, I'll run through the relevant concepts. You will also learn how two coils on a ferrite rod can act as a **transformer**.

Then, if you want to hear the sound through a speaker instead of in an earphone, you'll need to boost the signal some more. The easiest way to achieve this is by adding a simple integrated circuit chip known as an LM386.

By the end of this experiment, you will have built AMR2, a "real radio" capable of bringing

You Will Need:

Keep the components in the final configuration that you used at the end of Experiment 1. You will still need the same battery, connection blocks, speaker, ferrite rod, variable capacitor, and earphone, plus the resistors, capacitors, and 7555 timers from that experiment.

The list below is in addition to the components in Experiment 1:
- 26-gauge solid-core hookup wire, any color. (This time, you will need 15 feet.)
- Generic red LED (1).
- 2N3904 bipolar NPN transistors (3).
- LM386 amplifier chip (1). Made by Texas Instruments or National Semiconductor (now also Texas Instruments).
- Trimmer potentiometer, 10K (1).
- Resistors: 22 ohms (1), 100 ohms (2), 2.2K (2), 6.8K (2), 10K (2), 47K (5).
- Ceramic capacitors: 4.7nF (1), 10nF (5), 47nF (1), 0.1µF (3).
- Electrolytic capacitors: 10µF (2), 100µF (1), 470µF (2).
- Small piece of card stock, such as index card, measuring 2.5"×1.5".
- Clear adhesive tape.

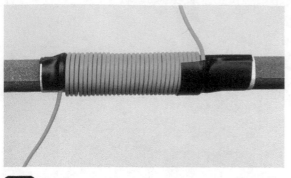

2-1 *The first layer of wire consists of 31 turns, secured with tape at both ends.*

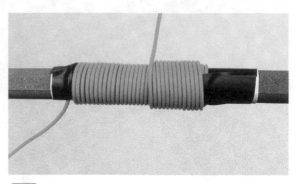

2-2 *Adding the second layer.*

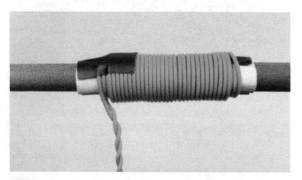

2-3 *The first coil is now complete.*

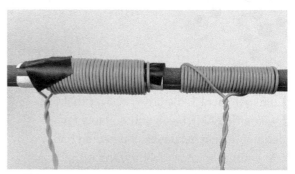

2-4 *The second coil is wrapped directly around the ferrite rod, as this coil will not be sliding.*

in stations from as far as 200 miles away. What will you hear? You'll have to tune in to the AM waveband to find out. The radio spectrum is full of surprises!

A DUAL-COIL DESIGN

AMR2 will eliminate the need for an antenna wire and a ground wire. The same ferrite rod that you used previously will now be your antenna, and transistor amplification will compensate for its smaller output.

You will need two separate coils, which will interact with each other. To understand how this works, you can perform a quick and simple test.

First, remove the wire that you wrapped around the ferrite rod in the previous experiment. You will need a larger coil now, consisting of 63 turns of 26-gauge wire. To keep the coil from taking up too much space, you'll be creating it in two layers. You will find that 10 feet of wire should be sufficient for 63 turns, and if necessary, you can make the coil from separate lengths of wire with their stripped ends twisted together.

Because you will want to slide the coil along the rod, you must wrap it around a new sleeve of thin card, similar to the one you made before. This time, the piece of card stock should be 2.5" × 1.5", and you will roll it around the ferrite rod with the long edge of the card parallel with the rod. Secure it with tape, allowing just enough looseness for it to slide along the rod, as before.

Allow a 6" tail to hang down at the beginning, and use a 1/2" square of tape to anchor the wire to the card. Now, start winding your turns as closely as possible. When you have completed 32 of them, add another piece of tape to stop them from unwinding. This is shown in Figure **2-1**.

Now, continue winding the wire in the same direction, but in a new layer on top of the first, returning toward your starting point, as shown in Figure **2-2**. Finish up with another piece of tape, as in Figure **2-3**.

For our dual-coil demonstration, you now need a second coil consisting of 31 turns of 26-gauge wire. This is wrapped tightly around the same rod, in the same direction as the first coil, but as near as possible to the opposite end, as shown in Figure **2-4**. You won't be sliding it, so it does not require a sleeve of card stock. After you create it, if you twist the insulated ends together, you shouldn't need any tape to hold it in place. Five feet of wire will be sufficient, including a 6" tail at each end.

Now, you can use your test circuit from the previous experiment to demonstrate how one coil can communicate with another. The interaction works better at a high frequency, so you'll only need IC2, not IC1. For clarity, we've shown its little circuit on its own in Figure **2-5**. You can compare this with Figure 1-10, which shows IC1 as well as IC2. But IC1 will not be used and should

2-5 *A test circuit that will demonstrate one coil communicating with the other.*

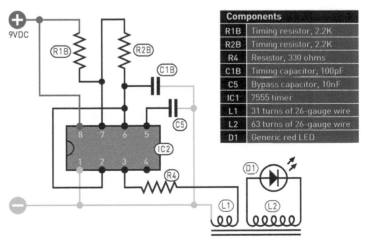

Components	
R1B	Timing resistor, 2.2K
R2B	Timing resistor, 2.2K
R4	Resistor, 330 ohms
C1B	Timing capacitor, 100pF
C5	Bypass capacitor, 10nF
IC1	7555 timer
L1	31 turns of 26-gauge wire
L2	63 turns of 26-gauge wire
D1	Generic red LED

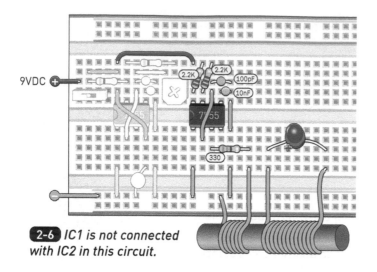

9VDC

2-6 **2-6** *IC1 is not connected with IC2 in this circuit.*

not be connected with IC2. Therefore, disconnect the yellow wire that went to Pin 4 of IC2. This is shown in Figure **2-6**.

You must also remove the yellow loop antenna from Experiment 1 (temporarily, at least).

Because IC1 will not be used right away, and has been disconnected from IC2, I am only showing the revised part of the circuit in Figures 2-5 and 2-6. The shorter (31-turn) coil on the ferrite rod has been inserted where the loop antenna used to be, and the longer (63-turn) coil is connected only with an LED.

The LED can be inserted in the board either way around because it will block half of the oscillations, just as the diode did in Experiment 1. There won't be enough voltage to damage the LED when it is blocking the current.

If you slide your 63-turn coil up and down the rod, the LED will get brighter and dimmer depending on the interaction of the second coil with the first coil. One coil on the rod is *inducing* a voltage and current in the other coil, with enough power to light the LED.

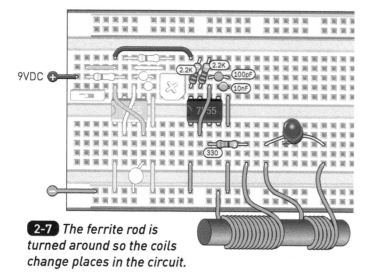

9VDC

2-7 *The ferrite rod is turned around so the coils change places in the circuit.*

Now, disconnect the coils, turn the ferrite rod around, and reconnect the coils, as shown in Figure **2-7**. The 63-turn coil is now attached to the output from IC2, while the 31-turn coil powers the LED. And—nothing happens!

The first configuration supplied just enough voltage for the LED, while the opposite configuration couldn't deliver the minimum voltage that an LED requires. Unfortunately, you can't verify this with a multimeter, because the meter won't respond to such a high-frequency voltage, even when you set it to measure AC. Still, the LED tells you all you need to know.

TRANSFORMER MATH

Whichever coil you attach to the output of IC2 is the **primary** coil in this experiment, and whichever coil drives the LED is the **secondary** coil. We will use some abbreviations to write a formula that you may want to refer to later:

- **TP:** turns of wire in the primary
- **TS:** turns of wire in the secondary
- **VP:** voltage applied to the primary
- **VS:** voltage measured in the secondary

And now, the formula:

TP / TS = VP / VS

In other words, if the number of turns in the primary relative to the secondary is 1:2, as in Figure 2-6, the voltages have the same ratio. Twice as many turns in the secondary should induce twice as much voltage. Does this mean you are getting something for nothing? No, because there's another formula telling you how the current behaves.

- **IP:** current passing through the primary
- **IS:** current drawn from the secondary

TP / TS = IS / IP

In other words, you may get twice as much voltage, but you can only draw half as much current. The overall power supplied to the primary coil is the same as the power taken from the secondary coil, because power can be defined as voltage multiplied by current.

Wait a minute, though. In the experiment that you just performed, each pulse applied to the primary will be around 9V, since the timer has a 9V power supply. So, with the 31-turn coil attached to the output from the timer, the secondary should have delivered 18V. Why didn't it vaporize your LED?

Because the formula only gives you twice the voltage in theory. In reality, the conversion is much less than 100% efficient. Even if you used dozens or hundreds of turns of tightly packed magnet wire, you would not reach 100% efficiency. A few dozen turns of hookup wire are maybe only 5% efficient— but they are enough to provide this take-home message:

A 1:2 ratio of turns in the primary coil compared with the secondary will deliver enough voltage to make the LED glow.

A 2:1 ratio does not.

TRANSFORMING POWER

When you have two coils sharing a common core, and they step up (or step down) the voltage, they function as a ***transformer***.

The world as we know it couldn't work without transformers. Electricity is distributed through long-distance power lines at 25,000V or even higher. Then this voltage is stepped down, and down again, until the power distributed through wires on poles in rural areas may be at 500V, at which point one final transformer delivers 220V to a residence.

This system is used because it saves money. Perhaps you remember Ohm's law. If **V** represents voltage measured in volts, **i** represents current in amps and **R** represents resistance in ohms:

$$V = i * R$$

Perhaps you also remember the formula for power. If **W** represents power in watts, **V** represents volts and **i** represents amps:

$$W = i * V$$

In the second formula, we can remove **V** and substitute **i** ***** **R** from the first formula:

$$W = i * i * R$$

If electric power is transported with a wire with resistance R, the power W lost in the wire is proportional to the square of the current. This power is lost as heat. Evidently, it would be a good idea to transmit power with as little current as possible. How can you do that?

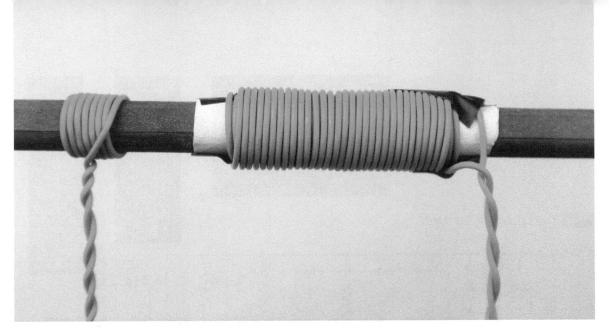

2-8 *The coil of 31 turns has been replaced with a coil of 7 turns.*

The obvious way is to increase the voltage.

So much for my digression into transformer theory. Now, why would you need a transformer in a radio receiver? Because the current output from the coil in your antenna has to be high enough to satisfy the requirements of the transistors we are going to use to amplify the signal.

AMR2 SETUP

You can keep the first coil that you just created, of 63 turns, but the smaller coil of 31 turns is the wrong size for the radio. Unwind it and reuse some of the wire to make a little coil of only 7 turns, near one end of the rod, as shown in Figure **2-8**. The coil of 63 turns must still be able to slide to and fro. (It does not matter if the turns in the coils are both in the same direction.)

The small coil can be movable as well, if it's not wound too tightly. The idea is that you can slide the large coil to set the tuning range and then move the small coil to optimize the signal strength. For weak signals, moving the small coil right next to the large coil increases the signal strength at the expense of selectivity; for strong signals, moving the small coil away increases the selectivity by disturbing the resonance circuit less.

With a coil-turns ratio of 63:7, you can see that if the big coil is your

Other Components	
L1	7 turns of 26-gauge wire
L2	63 turns of 26-gauge wire
VC1	Variable capacitor, 200pF
D1	Generic red LED
Q1,2,3	2N3904 bipolar transistor
P2	Trimmer potentiometer, 10K
LS1	High-impedance earphone

Capacitors	
C5	100µF
C6	0.1µF
C7	10nF
C8	10nF
C9	10nF
C10	10nF
C11	10nF
C12	0.1µF
C13	10uF
C14	4.7nF

Resistors	
R4	100
R5	10K
R6	47K
R7	6.8K
R8	2.2K
R9	47K
R10	6.8K
R11	2.2K
R12	47K
R13	47K
R14	47K
R15	10K

2-9 *The schematic for AMR2.*

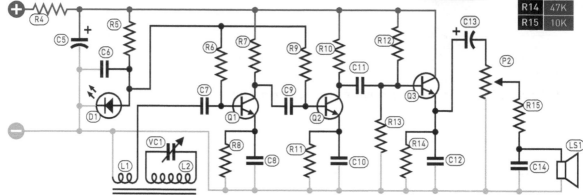

antenna and the output from the little coil goes to a transistor amplifier, this arrangement will boost the current by a factor of 9:1, which will be necessary for transistors in the circuit.

The schematic for AMR2 is shown in Figure **2-9**. The test circuit consisting of the two timers is not shown in this schematic, but it should still be on the breadboard because you'll be using it in a moment. Figure 2-9 only shows the new, additional components because they are not connected electrically with the old components. The breadboard layout is shown in Figure **2-10**, with the test circuit grayed out.

Note that the loop antenna has been put back where it was in Figure 1-9, together with the jumper from Pin 3 of IC1 to Pin 4 of IC2. The two 7555 timers are now ready to transmit a test signal from the loop to your ferrite rod, which will function as the receiving antenna.

Now you can see why we didn't use the positive bus on the breadboard previously. We were saving it for AMR2. (I am assuming that your breadboard

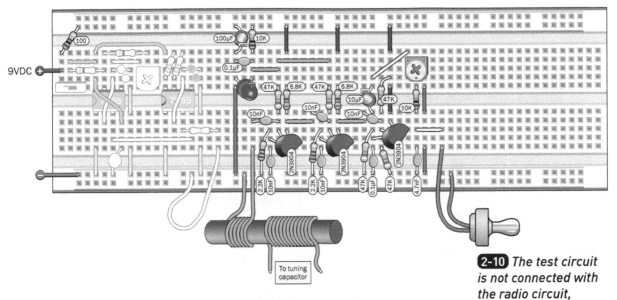

2-10 *The test circuit is not connected with the radio circuit, because the 63-turn coil on the ferrite rod will pick it up.*

has positive and negative buses that extend from end to end of the board without a break in the middle. If the connection in the buses does not extend from end to end, you will need to add jumper wires in the middle.)

Notice the 100-ohm resistor at top left, labeled R4 in Figure 2-9. This supplies power through the bus to the new components in the middle of the board. R4 works with C5 to provide filtered power to the transistors to prevent them from oscillating. In an amplifier circuit, oscillation can occur if the output can feed back to the input, even through the power supply. Remember:

Don't omit resistor R4!

Because the 63-turn coil (identified as L2) is now being used as the antenna, it is wired in parallel with the 200pF variable capacitor (identified as VC1) to adjust the resonance of the circuit when you are searching for radio stations. L2 is not connected with the breadboard; it just steps up the current through L1, which connects through a small coupling capacitor to Q1, the first transistor. Q1 begins the process of amplifying the radio signals.

L2 must be connected with the tuning capacitor, as shown in Figure **2-11** on the following page. I assume you will disconnect the tuning capacitor from its previous role in Experiment 1. You can connect it now through the three-position piece of terminal block that you made previously.

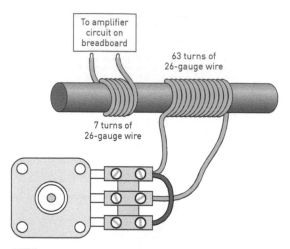

2-11 *You need to add the capacitor to adjust the resonance of the coil.*

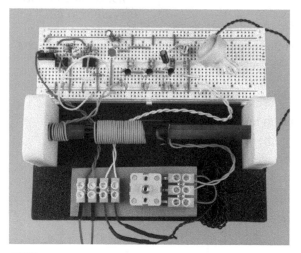

2-12 *The finished project with components mounted on squares of plastic to make them easier to handle.*

You may want to mount the ferrite rod, the tuning capacitor, and the breadboard on a piece of plastic or plywood to anchor them securely. Otherwise, if the rod and the capacitor slide around, they are likely to pull wires out of the breadboard. In Figure **2-12**, an additional block consisting of four pairs of terminals has been added at lower left to deal with the power supply and the output to the earphone.

TESTING AMR2

Don't expect to power AMR2 from an everyday AC adapter. If you try, you'll most likely hear a shrieking sound, as adapters tend to have a noisy output that will be picked up by the ferrite-rod antenna and amplified by the transistors in this circuit. When you are testing AMR2, always start by using a 9V battery. Fortunately, the radio doesn't use much power; it should run continuously for a couple of days from one battery.

Apply the battery voltage to the end of the board, as shown in Figure 2-10. Turn the trimmer potentiometer P2, shown in Figure 2-9, all the way counterclockwise. Now, use the slide switch at the left end of the board to power up the two 7555 timers. Make sure the output of IC1 (from Pin 3) connects to the reset pin of IC2 (Pin 4), while the output of IC2 (from Pin 3) is connected through the wire loop antenna.

Insert the earphone in your ear, and turn P2 clockwise. This is your volume control. You should hear the test tone, which is being picked up by L2, the larger coil. Adjust P1, and you should detect the full range of audio test frequencies. Adjust the tuning capacitor (identified as C15), and you should hear the test tone strongly at the point where the receiver is tuned to the same frequency as IC2.

If this all works the way it should, switch off the test circuit, which you won't need anymore. You can turn the volume all the way up and start searching for radio stations.

During the daytime, you may not find many. About an hour after sunset, everything starts to happen. If you turn the tuning capacitor slowly, you're likely to pick up at least two or three stations. Bear in mind that the performance of the ferrite rod is directional. For best results, the rod should be 90 degrees to the direction from which a radio broadcast is coming to you.

Three or four hours after sunset, your radio reaches its maximum performance level. You may find that some stations start to fade slowly in and out. This is because a distant signal may reach you via multiple paths as it bounces off an upper level of the atmosphere known as the Heaviside layer. (See page ix.) When one path is slightly longer than another, the phases of the transmission can cancel. Then the conditions in the atmosphere change slightly, and the effect disappears, allowing the sound of the radio station to return. This is part of the AM listening experience, as signals from distant sources struggle to find their way to your antenna. Maybe this seems like a disadvantage compared with FM, but FM transmissions have a much shorter range of only 30 to 40 miles. AM can reach you from 10 times that distance, depending on the power of the transmitter. A sophisticated AM communications receiver with a suitable antenna can pick up radio stations in the medium waveband from more than 1,000 miles away.

IF IT DOESN'T WORK

First, wait till after dark. In a remote location, you may hear no radio stations at all during daylight hours (which is why we included the test circuit to confirm that your radio is functional).

Second, remember to switch off sources of interference in your location. Now that you are amplifying radio signals, interference will be amplified, too. In fact, it can drown radio signals out. Even after an LED desk lamp is switched off, the transformer-rectifier that powers it may still create some hiss or buzz. You may need to unplug devices of that kind completely if they are close to AMR2.

Fluorescent lights are a bad source of noise—even ring lights with a magnifying glass in the center, popular among people assembling small

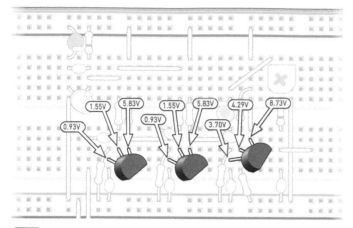

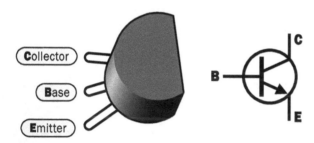

2-13 *These are the approximate voltages (relative to negative ground) that you should find if you made no wiring errors.*

2-14 *Names of the leads on your 2N3904 transistors.*

parts on a workbench. Decorative electroluminescent lights can also create interference.

Third, if the signal in your earphone isn't loud enough, remove R15 and substitute a jumper wire.

When searching for radio stations, begin with L2 as close as possible to L1 on the ferrite rod, in exactly the same way as when you were using two coils to illuminate an LED. This stronger coupling provides a stronger signal to the radio. Unfortunately, stronger coupling makes the resonance circuit less selective, so stations may seem to overlap. In general, you want the weakest coupling that gives adequate signal strength, to keep the selectivity as high as possible. If you move L2, you may have to readjust VC1 to compensate.

What if you hear nothing at all—not even the signal from your test generator? You probably made a wiring error, but you should also look for bad connections between jumpers and the metal clips inside your breadboard. (It is a sad fact that as breadboards have dropped in price over the years, their quality seems to have diminished.) Set your meter to measure DC volts, and attach the black lead to negative ground on the breadboard (using alligator test leads if necessary). Then check voltages on the leads of your transistors. Figure **2-13** shows the DC voltages that I found, using a power supply of 8.92V. If your values differ significantly, you may have made a wiring error, or a resistor value may be wrong.

You can also test your transistors using your meter, remembering that *C* identifies the collector, *B* is the base, and *E* is the emitter. Figure **2-14** identifies the leads on a 2N3904 transistor, while Figure **2-15** shows this type of transistor being tested. This can be confusing, as you may have to turn the transistor around to match the leads correctly.

Remember to set your meter to test for an NPN (not-PNP) transistor. If you insert the leads in the right holes, you should see a reading of around 220 on the meter display. This is the amplification capability of the transistor, also known as its beta value. If your meter beeps, or you get an error message or a value below 100, either you have inserted the transistor incorrectly or it has been damaged somehow.

Note that the LED in this circuit is not just a power indicator. Together with R5 and C6, its main function is to provide a constant voltage of around 1.6V for the bases of the transistor

2-15 *Testing a 2N3904 using a meter.*

amplifiers. While the circuit was running, I measured a voltage of 1.68V on the right-hand lead of the LED, which is marked with a + sign in Figure 2-10. If you insert the LED the wrong way around (with its long lead on the right instead of on the left), the circuit won't work.

What if you do hear radio stations, but they are all at one end of the range of your tuning capacitor? Try removing it from the terminal block and adjusting the tiny trimmer capacitors underneath (the two screws shown in Figure 1-21). You can also change the number of turns of wire on the ferrite rod. The inductance of each coil is proportional with the number of turns squared, which has been chosen to satisfy the range of frequencies in the medium-wave AM broadcast band, but this may not be optimal for your particular ferrite rod and the wire that you chose to create coils. If you have difficulty hearing stations at the low end of the AM frequency range, try removing some turns from the coil.

Bear in mind that the circuit does not have a power amplifier—yet! We'll add that in the next and final step in this experiment. That may make a big difference in your ability to pick up radio stations.

HOW IT WORKS

If you look at the schematic in Figure 2-9, you'll see that the first two transistors, Q1 and Q2, act as voltage amplifiers and that the layout of components around each of them is identical. If you are thinking that Q1 applies some initial amplification and Q2 increases it further, you are correct.

The third transistor, Q3, acts as a peak detector and demodulates the AM signal. (We used a diode for this purpose in AMR1.)

The way in which Q1 and Q2 are wired is known as the **common-emitter configuration**. In each transistor, the emitter is shared with the base and the collector; it interacts with both of them. When the voltage applied to the base increases, the effective internal resistance of the transistor diminishes, and more current flows between the collector and the emitter. In this way, the transistor is a current amplifier, but let's consider the consequence of the reduction in effective resistance. The collector now has less resistance between itself and ground, so its voltage drops. But if the voltage on the base diminishes, the effective resistance of the transistor increases, and so does the voltage on its collector.

In this way, if output is taken from the collector, it has a higher voltage range than the range applied to the base.

Because this circuit is dealing with rapidly varying radio signals, coupling capacitors such as C7, C9, C11, and C13 are used, blocking DC and just allowing the fluctuations to get through—because they are what we want.

(The design of transistor amplifiers is a complicated topic. Whole books have been written about it. Because this is a book about radio, and I have limited space, I won't go deeper into it here.)

C14 and R15 form a low-pass filter that removes high frequencies from the audio signal before it reaches the earphone or the audio amplifier. This reduces noise and a particular high-pitched sound that may form when different AM stations interfere with each other. You can experiment with different values for C14, such as 2.2nF or 10nF, or try removing it completely.

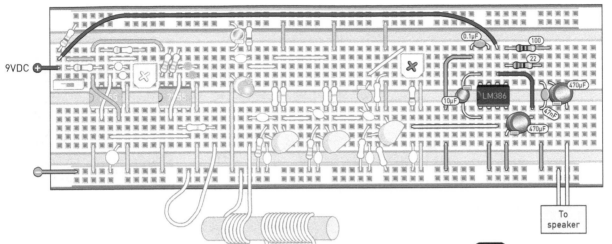

Labels on the figure: 9VDC, 0.1µF, 100, 22, 470µF, 470µF, 10µF, 47nF, LM386, To speaker

2-16 *The LM386 amplifier chip has been added, with a horizontal yellow wire connecting its input with the rest of the circuit.*

ADDING AUDIO POWER

Although the transistors that you installed in the circuit provide some amplification, they do not deliver enough power to drive a speaker. For that purpose, you can now complete the circuit of AMR2 by adding an LM386—an old-school chip that can amplify its input by a factor of 200.

Adding the LM386 is very easy: Just disconnect the earphone and add the new horizontal yellow wire visible on the right in Figure **2-16**.

Notice the long, red power-supply wire in this figure, which is necessary because the positive bus has been reserved for filtered power to the transistors. Pin 6 on the LM386 happens to be its power-supply pin, while Pin 5 is its output and Pin 3 is its input.

Figure **2-17**, on the following page, shows a schematic of how LM386 is added to the previous circuit. R17 and C18 in the schematic are optional; they bypass the speaker to reduce the amount of hiss you would normally hear in AM transmissions. You can experiment by removing these components. R18 prevents the speaker from drawing too much current. C19 protects the speaker from DC voltage. C17 removes fluctuations from the power supply. C15 is a bypass capacitor to remove noise.

An interesting feature of the LM386 is that it allows you to adjust its amplification ratio by inserting components between Pins 1 and 8. If these

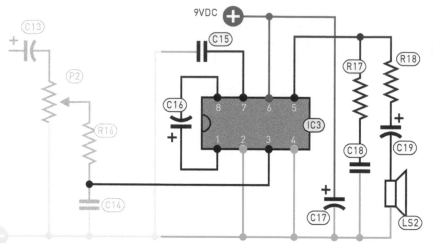

Additional Components	
C15	Ceramic 0.1uF
C16	Electrolytic 10uF
C17	Electrolytic 470uF
C18	Ceramic 47nF
C19	Electrolytic 470uF
R17	22 ohms
R18	100 ohms
IC3	LM386
LS2	Small speaker (8 ohms)

2-17 *The schematic version of the breadboard layout in Figure 2-16.*

pins are left unconnected, the default amplification factor is 20:1. If a 10μF capacitor is inserted (as shown in Figure 2-17), the amplification increases to 200:1. Intermediate values for C16, used in series with resistances, will allow different amplification values. In case this interests you, the datasheet for the LM386 supplies more details and some examples of circuits.

SOURCES AND FURTHER READING

For the three-transistor design of AMR2, I was inspired by two sources: the HJW Electronics Breadboard Six Transistor Radio BBRK-1, by Henry J. Walmsley, which is at

www.usefulcomponents.com/main_contents/projects/breadboard_trf _radio/breadboard_trf_radio.html,

and the book *Build Your Own Transistor Radios*, by Ronald Quan. For more theory and design process of transistor amplifiers, I recommend additional sources: *Practical Electronics for Inventors*, by Paul Scherz and Simon Monk, has a short but good section on transistor amplifiers. *The Art of Electronics*, by Paul Horowitz and Winfield Hill, also contains the required theory, but it's a bit more spread out.

For online resources, I recommend www.electronics-tutorials.ws/amplifier/amp_2.html.

This contains several amplifier circuits and the theory of how they work.

The Bipolar Transistor Cookbook, which has been published in multiple parts at the website Nuts and Volts, contains amplifiers, among many other transistor circuits. See www.nutsvolts.com/magazine/article/bipolar_transistor_cookbook_part_3.

STATUS UPDATE

This section began by explaining how coils can interact to modify current and voltage. You also saw how one coil can transmit enough power to another coil to light an LED. This basic principle is used in transformers throughout the power grid delivering electricity to your home—and in radios, where a signal has to be transformed to suit the needs of transistors. In AMR2, you saw how bipolar transistors can boost a radio signal so that it becomes powerful enough to drive a speaker.

Now that you have a functional radio, it's time for the next step: Build a transmitter that can broadcast music, voice, and other sounds, on a limited scale, that AMR2 will be able to receive.

Experiment

3

A REAL AM
TRANSMITTER

The transmitter that you built in Experiment 1 was a bare-bones device that could only send test tones. You can think of it as being AMTzero.

Now it's time for AMT1, which is built around only one transistor, yet can send speech and music. You will be able to pick up its signal with an off-the-shelf radio, or you can receive it using AMR2, the receiver that you built in Experiment 2 (so long as you don't need to repurpose those components for this circuit).

The power of a homemade transmitter is limited by regulations that vary from one country to another. In this experiment, it's also limited by using a single transistor. Still, if you've ever had the desire to serve as your family deejay, broadcasting radio programs that other people can tune in to at home, this device is a step toward that goal. It's a small step, but it demonstrates the possibilities.

COMPONENT NOTES

A note about inductors: An inductor looks like a resistor, but it contains a tiny coil. Remember that *H* stands for "henries." Henries are the units of inductance, and *µH* means "microhenries." Inductors are also rated for the maximum continuous current that they can pass. You may

You Will Need:

For this experiment, you will need some new components, which you can find described in Appendix A and Appendix C:

- Inductor, 22µH, maximum 40mA (1).
- Audio cable, at least 3 feet long, with a 1/8" audio plug at each end (1).
- Audio-to-breadboard adapter (1). This consists of a 1/8" audio socket (also known as an audio jack), connecting with pins that can plug into a breadboard, or screw terminals that can be wired to a breadboard.
- Any audio device that has a headphone output from a 1/8" socket (1).

If you are willing to disassemble the circuits that you built in Experiments 1 and 2, you can reuse those components, and you will just need a few extras:

- Resistors: 1K (3), 4.7K (2).
- Ceramic capacitors: 2.2nF (1), 47nF (1), 1µF (1).
- AM radio receiver (1). I listed this as an option before, but if you are going to disassemble AMR2, you will need some other way to receive the AM radio signals you will be transmitting, and a handheld portable radio is the easiest option.

If you are *not* willing to disassemble the previous circuit and reuse the components, you can ignore the list above. You can use AMR2 as your radio receiver, but you will need these components for AMT1:

- A second breadboard (1).
- An additional tuning capacitor, 200pF, type 223P (1).
- Resistors: 1K (3), 4.7K (2), 10K (1).
- Ceramic capacitors: 2.2nF (1), 47nF (1), 0.1µF (1), 1µF (1).
- Electrolytic capacitor, 10µF (1).
- 2N3904 bipolar NPN transistor (1).
- Extra hookup wire, either 22-gauge or 26-gauge, to make a 4" loop antenna (2 feet).

LEGAL STUFF

Because the AM radio spectrum is a limited resource, government agencies like to ration it. Depending on the country that you live in, you may find there are strict rules limiting the power and range of any transmitter that you build.

In the United States, a homemade transmitter for the AM band is allowed to broadcast with a power of up to 100mW, which is minuscule compared with the 50,000 watts of some "real" radio stations. Also, your antenna (including the wire to it, and any wire to ground) may be no longer than 10 feet.

If you want to know all the details, search online for
FCC regulations "part 15"

Our circuit for AMT1 only uses a small loop antenna, and powers it with much less than 100mW, so we think it should be totally legal in the United States. In some other countries, the rules may be more strict. I cannot check them all, so I suggest you search online if you are concerned about this issue.

Of course, there is nothing to stop you from buying components and putting them together to create any transmitter if you wish. Your problems only begin if you switch it on and use it, especially if you interfere with other people's radio reception. I'll show you how to minimize the risk of this happening, but I take no responsibility if you decide to push the limits. That is up to you.

substitute a higher current rating, but not a lower current rating, than the one I specify.

A note about audio sources: You can use almost any device that has a headphone jack to serve as an audio source for the transmitter in this project. A laptop computer, an MP3 player, a boombox, or (maybe) an old mobile phone can serve this purpose. Many modern devices don't have headphone jacks anymore; you'll have to check. If you pick something up cheaply from a yard sale, such as a CD player or even an ancient cassette player, that will be ideal.

CREATING A CARRIER

In Experiment 1, I mentioned that a transmitter needs a carrier wave of a frequency between 530kHz and 1,700kHz to be classified as a radio operating in the AM band. We used a 7555 timer to generate a carrier, but it was a square wave, which is not ideal. Really, a carrier should be a *sine wave*. I'll explain this distinction in detail in a moment, but for now, here are the most important basics:

- An LC circuit can function as an *oscillator* and can generate a sine wave.
- An LC circuit can also detect a sine wave by resonating with it.

Turn back to Figure 1-40, which shows a sine wave that was created in an LC circuit by an initial voltage pulse. To use it as a carrier, we need to sustain it with just enough energy to keep it going indefinitely. How can this be done? A transistor wired as an amplifier can do it.

If you're wondering how this works, consider a microphone wired to a cheap speaker system.

If too much sound from the speakers gets back to the microphone, you hear a howling, whistling, or screeching sound commonly referred to as *acoustic feedback*. That's the kind of thing we want, although the frequency must be adjustable, and we need a way to keep the oscillations from getting out of control.

In this experiment, I am assuming you are starting with a clean breadboard, either by using a new one or by removing all the old components that you placed on a board in Experiments 1 and 2.

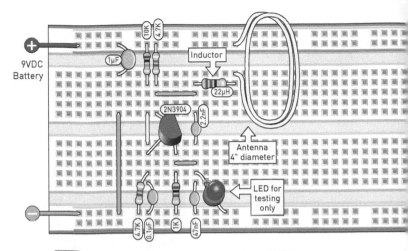

3-1 *The first step in breadboarding AMT1.*

Figure **3-1** shows the breadboard layout for a simple oscillator circuit that will be the basis for AMT1, and Figure **3-2** shows the schematic. The transmitting antenna is a loop consisting of two turns of 22-gauge or 26-gauge wire, about 4" in diameter. You will need slightly more than 2 feet of wire for this. You can use pieces of tape to hold the turns together.

Power is supplied by a 9V battery. When you built AMR2, I mentioned that an AC adapter will tend to introduce noise that interferes with radio reception. An AC adapter will also interfere with radio transmission, unless its output is very thoroughly smoothed.

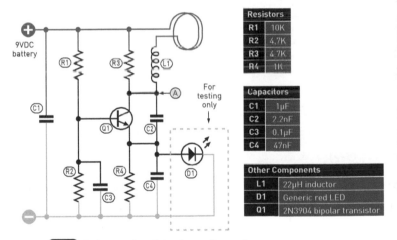

3-2 *Schematic of the breadboard layout in Figure 3-1.*

When you look at the circuit, you may notice that there's no input. That's because at this point, the circuit will simply generate a carrier wave. An audio input will be added in the next step.

You will also see that there is no audible or visible output from this circuit, other than the antenna, which transmits a carrier wave at almost 1MHz. So how will you know if it's working?

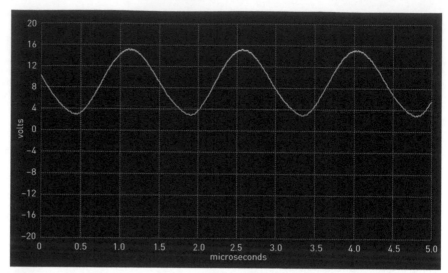

3-3 *The carrier wave generated by AMT1, as shown on an oscilloscope.*

If you have an oscilloscope, connect it between point A in the schematic and negative ground. You should see a nice, smooth sine wave such as the one shown in Figure **3-3**.

If you don't have an oscilloscope, you can test the circuit in two other ways. First, include an LED, as suggested in the breadboard diagram and the schematic. The LED should light up when you apply power to the circuit. This does not guarantee that the circuit is working, but it provides some reassurance.

Now, remove the LED. While power is applied to AMT1, switch on AMR1, AMR2, or any off-the-shelf AM radio receiver. If you have any sources of radio interference nearby, you should switch them off.

Hold your radio close to the loop antenna, and turn the tuning capacitor very slowly from 650kHz through 750kHz. As you turn the dial, you should find that the hiss of static that you normally hear from an AM radio disappears, and the radio goes silent. At this point, you are tuning into your transmitter's carrier wave. There's no sound, because we haven't added audio yet.

While your radio is tuned to the silent spot, disconnect power to the transmitter. Now, the static from the radio speaker resumes. Reconnect power to your circuit, and the static ceases again.

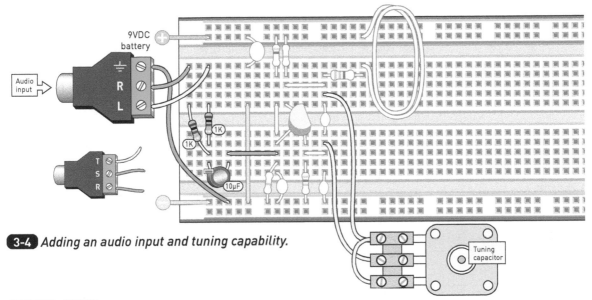

3-4 *Adding an audio input and tuning capability.*

ADDING AUDIO

The next step is to add an audio input and a tuning capacitor, as shown in Figures **3-4** and **3-5**. The tuning capacitor is the same type that you used before, and we are adding it so that you can adjust your transmitting frequency just in case there happens to be a radio station near where you live whose frequency you want to avoid.

The component at the left end of the breadboard is an adapter in which an audio jack is connected with three screw terminals. Most adapters have their terminals labeled with *L* (left channel), *R* (right channel), and the ground symbol. *A* few adapters have screws labeled *T*, *S*, and *R* (meaning tip, sleeve, and ring on the jack plug). Those screws have the same function as L, ground, and R. Figure 3-4 shows this type of adapter with the same color coding of wires as the more common type.

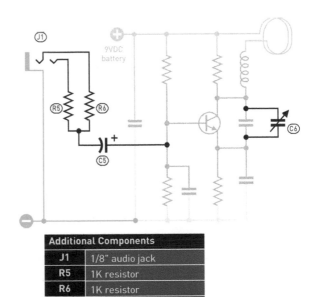

Additional Components	
J1	1/8" audio jack
R5	1K resistor
R6	1K resistor
C5	10µF capacitor
C6	Tuning capacitor, 200pF

3-5 *Schematic of the breadboard layout in Figure 3-4.*

Using an audio cable with a 1/8" plug at each end, connect an audio source with the adapter. The source is likely to be in stereo, and the plug at the end of your audio cable almost certainly will be a stereo plug. Because our transmitter only works in mono, we are using R5 and R6 to tie the two channels together. Then, the rapid fluctuations of the audio signal pass through C5 to the base of Q1.

Is your audio source safe when it is connected to this circuit? Naturally, I took steps to minimize any risk:

- The circuit is only powered with a 9V battery.
- C5 isolates your audio device from the DC voltage in AMT1.
- As a precaution, before you plug in your audio cable, you can insert your multimeter between the battery and the breadboard to measure the current consumption. If it is much more than 3mA, this may indicate that you made an error in the circuit.
- Headphone outputs on audio devices tend to be robust compared with more modern interfaces, such as USB ports.
- The 1K resistors connected to the audio input of our device offer additional protection.

When your audio cable is plugged into the headphone output of a boombox or other music player, this usually suppresses the speaker output from the player. Apply power to AMT1, switch on your radio, and position it as you did before. Tweak the tuning capacitor, if necessary, and the radio should pick up the music that you are broadcasting.

Most AM radios contain a ferrite-rod antenna, which is directional. You can turn it to optimize its reception. You can also turn up the audio source to create a more powerful input, but beyond a certain point, the music becomes distorted. This is known as *clipping*, and it occurs when your input exceeds the amplification ability of the transistor.

Try moving the radio farther away to find out how much range you can get. I was able to pick up the signal from a distance of 6 feet. That doesn't sound like much—but this transmitter is only working with a single transistor and a 4" loop antenna. AMT1 is really just for demo purposes, to show the possibilities.

Don't forget to disconnect your battery when you've finished your test.

Now, I have some explaining to do, beginning with the concept of a sine wave.

SINES

More than 2,000 years ago, ancient people divided a circle into 360 degrees, perhaps because each year contains about 360 days.

A quarter-circle has an angle at the center, which is 90 degrees, often referred to as a *right angle*. If you allow the two sides of a triangle adjacent to a right angle to vary in length, it could look like the one shown in Figure 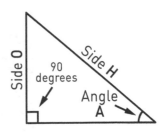. What if you want to know the size of angle A? Well, you could define it by measuring two sides of the triangle very carefully, such as side **O** (opposite to the angle) and side **H** (the hypotenuse of the triangle; this word is not used too often anymore but is still the correct term). Then divide **O** / **H**. This fraction is called the **sine** of the angle, which defines it uniquely. In mathematics, you would write:

sin(A) = O / H

(The abbreviation **sin** is universally used in formulas of this type.)

The fraction of **O** divided by **H** is a number that ranges from almost zero (when the angle is very small and side **O** is very short) to almost 1 (when the angle is almost 90 degrees and side **O** is almost as long as side **H**).

Now, imagine that an object tied to the end of a string is orbiting a point, as in Figure **3-7**. We can draw a series of triangles in which **O**, shown as red lines, varies depending on the angle at the center. Figure 3-7 shows these values at intervals of 15 degrees, because 15 divides neatly into 360. You can see that **O** starts from zero and gets longer as the angles increase in size, and then it gets shorter again as the object rotates past the midpoint.

On the right-hand side of Figure 3-7, the values of side **O** have been rearranged at regular intervals, so you can think of the horizontal spacing

Sine of angle A = $\frac{O}{H}$

3-6 *How can angle A be measured?*

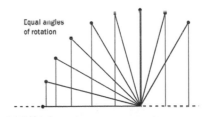

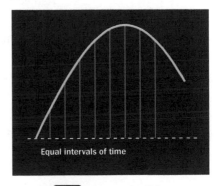

Equal intervals of time

3-7 *How to build a sine wave.*

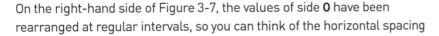

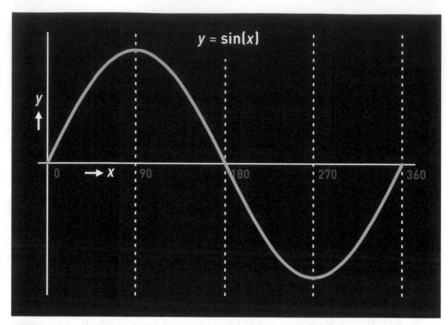

$y = \sin(x)$

3-8 *The way in which a sine wave is often shown in textbooks.*

as a timeline. Because the height of each line is proportional with the sine of the angle that created it, a smooth curve through the top ends of the lines is known as a **sine wave**.

In Figure **3-8**, one complete cycle of a sine wave is shown in the style that you will often find in textbooks, with a vertical axis named **y**, a horizontal axis named **x**, and the formula that creates the curve being

y = sin(x)

If you go to a search engine and enter this formula, you should see a sine wave on your screen. It may be divided into steps of 90, 180, 270, and 360 degrees, corresponding with the angles on the left side of Figure 3-7 when the object orbiting a point goes all the way around and returns to the place where it started.

Why does this matter? First, sine waves turn up all over the place:

- If you load a vertical extension spring with a heavy object and then allow the object to bounce at the end of the spring, you can plot its motion as a sine wave.

- The motion of a swinging pendulum can be expressed as a sine wave (so long as it doesn't swing too widely).
- If you take a soda bottle that is empty or almost empty and you blow slightly upward across the open neck of the bottle, the air inside it resonates, creating pressure waves you hear as a beautifully mellow tone. If you plot these rapid variations in air pressure over time, they form a sine wave.
- If you plotted the voltages of alternating current from an outlet in your house over a period of time, they would form a sine wave.
- The voltage in an LC circuit can be plotted as a sine wave, as shown in Figure 1-40.

The last item in this list is the one that matters to us here. Because an LC circuit naturally oscillates this way, you can use it in a receiver to resonate with a sine wave of a matching frequency. This leads to an important conclusion: An LC circuit in a radio receiver can detect a matching carrier frequency from a transmitter, while ignoring other frequencies, *so long as the carrier is a sine wave*.

This capability is really quite amazing when you consider the thousands or millions of radio signals in the electromagnetic spectrum. AM radio is just a tiny segment of this spectrum. Shortwave radio, FM radio, old-fashioned over-the-air TV transmissions, microwave transmissions, cellular telephone transmissions, satellite signals—the electronic cacophony continues day and night, all around you. But a simple circuit such as the one we used in AMR2 will filter everything out except a carrier wave in one narrow frequency band.

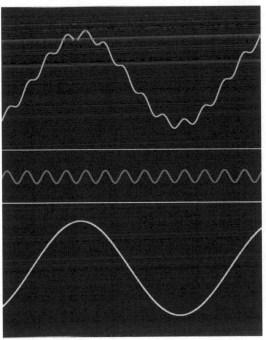

To get an idea of this process, understand that a complex wave such as the one at the top of Figure **3-9** is really the sum of the two waveforms shown below it. An LC circuit can extract either of them from the complex wave while blocking the other.

3-9 *An LC circuit should be able to extract either of the sine waves from the complex wave at the top.*

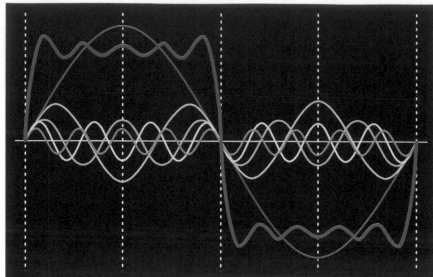

3-10 *It can be proved mathematically that a square wave, shown in red, can be created by combining many smaller sine waves, shown here in other colors.*

What if a transmitter creates a square wave? It's a strange fact that you can build a square wave by assembling a lot of sine waves of increasing frequency. This is suggested in Figure **3-10**, where the bold red curve is the result of adding all the other waves of different colors. The more curves you add of higher frequencies, the closer the red curve can get to being a square wave. (Of course, the size of each additional wave has to be just right, and there is a formula that defines this for you, although it's too complex for me to deal with here.)

The curves of higher frequency are known as the **harmonics** of the frequency of the square wave.

Because the LC circuit in a radio receiver resonates with sine waves, it would see a square wave as being made of multiple sine waves. Consequently, if you used a square wave carrier, it would be detected all over the radio spectrum, causing interference and confusion. We only used a square wave in our first demo because it was convenient with 7555 timers, and the transmitting power was so small that it would not cause any problems.

Here is the take-home message to remember:
• In the world of radio, all carrier waves are sine waves.

Now, you may be wondering: How does the simple circuit of AMT1 work to create the sine wave output?

UNDERSTANDING AMT1

First, recall how a transistor works: A voltage applied to its base controls current flowing in through the collector and out through the emitter—so long as the base voltage is at least 0.6V higher than the emitter voltage. Resistors R1 and R2 in Figure 3-2 provide a **bias voltage** to the base so that the base will remain in the active range.

R3 and R4 help to set suitable voltages on the collector and the emitter.

R3 provides an output voltage at point A, since the voltage at A depends on the current through R3 according to Ohm's law, and the transistor controls this current.

Now, consider capacitors C1, C2, and C4. If you follow the connections through them and around to the top of the circuit, you find that the path is completed through the antenna and the coil, L1. This is similar to but not quite the same as the basic LC circuit that we introduced in Figure 1-39. Capacitor C1 has a relatively high value, so it doesn't play a very active role, but it does allow high-frequency AC current to circle around.

The role of transistor Q1 is to sense the oscillation in the LC circuit (with the connection to the lower plate of C2) and increase the voltage to the upper plate of C2 at the right times to keep the oscillation going—like pushing a child on a swing.

You can think of this as a positive-feedback loop—just like feedback from loudspeakers to a microphone creating audio feedback in an auditorium.

Capacitor C3 on the transistor base has two effects: First, it stabilizes the base voltage and is necessary to keep the oscillator running. Second, by stabilizing the base voltage, it filters out high frequencies from the audio signal fed to the base. Together with R5 and R6 (which are shown in Figure 3-5), it forms a low-pass filter.

Technically, this is an example of a **common-base amplifier** circuit, because the base of the transistor is held at a constant voltage. It's a tricky circuit to understand because the input is really at the emitter, and fluctuations of the current flowing through the horizontal wire toward the emitter control the fluctuations in the coil connected to the collector.

This type of amplifier is not seen as often as a common-emitter amplifier of the type we used in AMR2, but one property makes it especially useful in the oscillator circuit: It is **noninverting**. A common-emitter amplifier inverts the signal; when the base voltage goes up, the output voltage at the collector goes down, and this makes it more complicated to arrange positive feedback required for oscillation.

The AMT1 circuit is also an example of a **Colpitts oscillator**, which has the distinguishing feature that one coil is used with two capacitors (C2 and C4) to create the resonance circuit, with feedback extracted from between the two capacitors.

If you look at the extended form of the circuit in Figure 3-5, you will see that the audio input is wired through capacitor C5 to the base of the transistor. Consequently, while the circuit is oscillating, the varying voltage on the base modulates the amplitude of high-frequency oscillations. (For complicated reasons, changing the base voltage in the common-base amplifier influences the **amplifier gain**, which is how much the amplifier amplifies a signal. We can use this fact to control the amplitude of the signal the oscillator produces.)

When I readjusted my oscilloscope to show frequency in kilohertz rather than megahertz, the output from the transmitter (measured at point A in the circuit in Figure 3-2) looked like the screen capture in Figure **3-11**. In fact, this particular trace was created by a recording of a bass guitar.

The oscilloscope does not really capture the oscillations accurately, because the sampling rate is too low. It is difficult to show AM signals with digital oscilloscopes because the time scales of the carrier wave and the audio frequencies are so different.

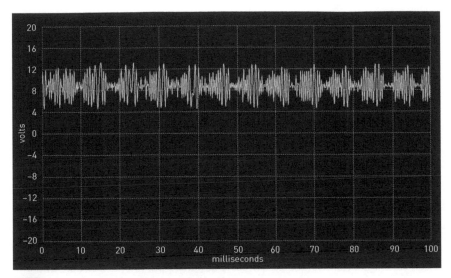

3-11 *An audio signal added to the output from AMT1.*

SOUND QUALITY AND OVERMODULATION

With no audio signal on the input, the oscillator outputs a sine-shaped signal with a constant amplitude, as in Figure 3-3. With an audio signal added, the amplitude of this signal varies with the momentary audio signal voltage: When the audio signal is positive the amplitude is larger, and when the audio signal is negative the amplitude is smaller. Increasing the amplitude of the audio signal at the source (perhaps it has a volume control) increases the size of the amplitude variations at the output.

What happens if you keep increasing the amplitude of the audio input? In particular, the negative parts of the audio signal turn out to be problematic because if the voltage at the transistor base is too low, the oscillator will stop completely. Then the output signal begins to look more like a square wave and is a problem because it decreases the sound quality of your transmission and causes interference at nearby frequencies.

Knowing when such overmodulation occurs can be difficult without an oscilloscope. I have chosen the component values in the circuit to avoid overmodulation with a line-out signal at full volume, but because outputs from audio equipment can be different, this may not be enough. Decreasing the volume on the audio source a little adds some safety margin.

INCREASING THE RANGE

If you use a loop antenna of larger diameter, or if it contains more turns of wire, you can increase the transmission range. Feel free to experiment, but remember that in the United States, 10 feet of wire is the limit, and other countries have their own regulations. Also bear in mind that changing the loop will change the transmitter frequency.

The frequency may also change if you have large magnetic objects near AMT1, or objects that can be magnetized. They can affect the inductance or capacitance of the loop.

Ideally, the transmission frequency would be set with an LC circuit inside a metal enclosure, protected from outside influence. Old radio equipment often contains little metal boxes around sensitive parts of the circuit, and in modern equipment, you can find them soldered to a circuit board.

For even better stability, you could use a quartz crystal instead of an LC circuit to set the frequency. Modern radio circuits almost always do this. Quartz crystals are also often used with microcontrollers to set the clock frequency accurately. The Raspberry Pi Pico we will use for our next experiment uses a 12MHz crystal from which it generates all clock signals it uses (specified to be accurate within 30 parts per million).

SOURCES

The LC oscillator here is derived from one found in *Build Your Own Transistor Radios*. You can read more about oscillators in *The Bipolar Transistor Cookbook*, Part 5, at Nutsvolts.com.

A more serious AM transmitter with explanation is shown at www.geojohn.org/Radios/MyRadios/AMXmitr/AMXmtr.html.

You can learn more about transistor theory, including how amplifier gain is affected by emitter current, in *Practical Electronics for Inventors*, by Paul Scherz and Simon Monk.

STATUS UPDATE

You've seen how to transmit and receive amplitude-modulated signals, and along the way, you have been introduced to concepts such as sine waves, harmonics, overmodulation, and the need for a coupling capacitor when connecting an audio source with an amplifier.

The range and quality of AMT1 leave a lot of room for improvement, but if you want to be a deejay, not only broadcasting music but also adding your own commentary, this can be easily done by using your computer as the source of music and mixing in your voice by plugging a headset into a USB port.

Next, I will introduce the Raspberry Pi Pico, a powerful microcontroller that turns out to have unexpected but important applications in radio.

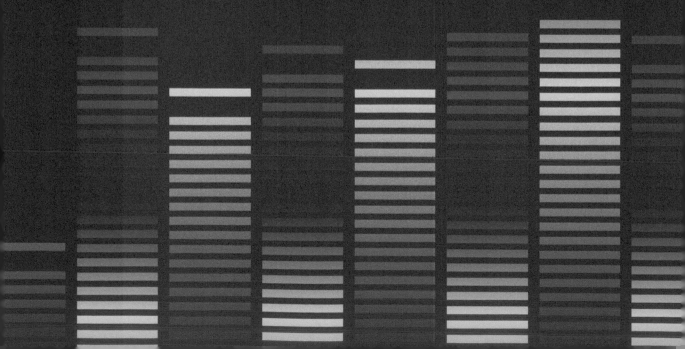

Experiment

4

RASPBERRY
PI PICO

A **microcontroller** is a programmable chip that controls other devices. What does that have to do with radio? The microcontroller that I am recommending may not have been designed primarily with radio in mind, but it's so powerful, it can operate fast enough to generate a carrier wave and can also measure frequencies. Its name is Raspberry Pi Pico.

In this experiment you will learn how to set up the Pico with your computer using any of the three major operating systems: Windows, Macintosh, or Linux. You'll find out how to use software known as an IDE, and I'll take you through some troubleshooting steps if everything does not work exactly as you expect.

You'll add a small LCD screen to the Pico that will display status messages and useful information such as radio frequencies. Then you'll install a program (sometimes referred to as a sketch). All the sketches in this book are easily accessed online, so you can open them from the IDE or paste them into its window.

By the end of this experiment, you'll be ready to enhance your experience of building radio receivers and transmitters.

You Will Need:

- Raspberry Pi Pico board (1). I recommend the genuine manufacturer's brand with the raspberry logo on it.
- USB cable with Type A plug on one end and Type B plug on the other end (1).
- LCD display model 1602 with I2C capability (1). Available from multiple manufacturers.
- *BAT48* Schottky diodes (2). This is the same type of diode you used in Experiment 1, but you now need two of them.
- Momentary switch (1). Also called a pushbutton.
- Jumper wires (at least 4, 6" in length, in assorted colors). Must be the flexible type with a connector at each end, and must be *male-to-female* type (socket at one end, plug at the other, as shown in Figure 4-3).

4-1 *Raspberry Pi Pico board with pins soldered in.*

4-2 *USB cable with Type-A plug on one end and Type-B Micro on the other end.*

4-3 *Flexible jumper with plug at one end and socket at other end.*

COMPONENT NOTES

The Pico is available in four formats: *Plain Pico* or *Pico W* wireless (with no header pins), and *Pico H* or *Pico WH* (with header pins soldered on). You must have header pins to plug the Pico board into your breadboard, so if you buy a Pico board without pins, you will have to buy the pins separately and solder them on yourself. You will only spend a couple of extra dollars for a Pico board with pins already attached, so I suggest that you do that. For our experiment, we don't need the wireless capabilities of the W or WH versions, so I suggest you get a Pico H.

The Pico connects with your computer using a USB cable with Type A and Type B plugs, as shown in Figure **4-2**. This has to be a data-transfer cable, not just a charging cable.

You will need flexible jumper wires that have a plug at one end and a socket on the other end, as shown in Figure **4-3**. Don't make the mistake of buying jumpers that have plugs at both ends.

GETTING SKETCHES

Your Pico needs instructions that you provide in the form of a program. This is known as a *sketch* among people who use the Arduino brand of microcontrollers, and I will be using that term here.

Writing a sketch requires some knowledge of a programming language, but if you don't want to tackle that learning curve, you can simply go online to find a sketch that someone else has written and upload it into the microcontroller. In fact, all the sketches that you need for this book can be found on GitHub, a website for developing and publishing software, at github.com/fjansson/MakeRadio.

After navigating to the above web page, follow these steps:
- Click the name of the folder that you want.
- Click the name of the sketch that you want.
- Press Control-A (Command-A on Mac) to select all the text.
- Press Control-C (Command-C on Mac) to copy the text.

After that, you will be able to paste the text into an application on your computer known as an *integrated development environment*, or *IDE*. This will upload the text as a sketch into your microcontroller via a USB cable.

Now, how do you get the IDE—the software which makes all this possible? I will take you through the necessary steps.

WHY THE PICO?

Released in 2021, the Pico is powerful but inexpensive. We will use it to control an FM receiver chip, to measure signal frequencies, and to generate signals at accurately known frequencies while displaying information on an LCD screen. The Pico is fast enough to generate signals in the AM broadcast band, and we will use it to make an AM transmitter with better sound quality than the one we built in Experiment 3.

Some readers may wonder, why don't we use an Arduino?

Compared with the low-end Arduinos (such as the Arduino UNO, which is based on the Atmel ATmega 328P microcontroller), the Pico is less expensive and much more powerful. There are other Arduino boards with more computational power and various features, but the large number of models is confusing, and I would have difficulty testing them all and determining which ones will do exactly what we want for our purposes here.

The Pico has quickly become popular among makers, and I believe it will be around for a long time. You can program it from your computer using the *Arduino IDE*, an interface that is very familiar to many people and allows us to use the huge variety of code libraries available online.

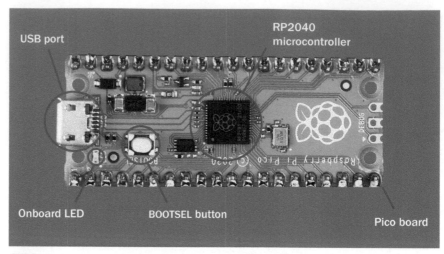

4-4 *Parts that I will be referring to on the Pico board.*

Multiple microcontroller boards are made by third parties using the same RP2040 chip developed by Raspberry Pi. My program code may work on these third-party boards, but again, I can't test them all, especially as new versions will have been released by the time you read this. Therefore, I suggest we restrict ourselves to the original Raspberry Pi Pico board branded with the raspberry logo.

TERMINOLOGY

The **Pico board** is the green rectangle with gold connectors on the edges. The **RP2040 Pico microcontroller** is the chip mounted in the center of the board, as shown in Figure **4-4**. When people talk about "the Pico," sometimes they are referring to the board, while other times they are talking about the microcontroller itself. I will try to distinguish carefully between the two.

The RP2040 uses ARM architecture. ARM processors of different kinds and with different features are found in most smartphones, in Raspberry Pi single-board computers, and even in the Japanese Fugaku computer, currently the fastest computer in the world. This chip is a 32-bit microcontroller, meaning its most basic operations use binary numbers with 32 digits. The microcontroller contains 264 kilobytes of RAM memory for storing variables and 2 megabytes of nonvolatile flash memory for storing a sketch. It runs at up to 133MHz and has two cores, meaning it can run two tasks in parallel. By microcontroller standards, this is powerful.

WHY THE ARDUINO IDE?

You can skip this section if the reasons for my choice of language and environment are uninteresting to you. I am including it in case some readers wonder about the choices I made.

First I had to choose which **programming language** to use. **C language** is old-school fast but not very user-friendly—while **Python** is more recently developed and intended to be easier than C. I also had to choose a **development environment**, which affects which code libraries are available to us.

I considered the following four options for programming the Pico:

C/C++ SDK

The acronym **SDK** stands for "**software-development kit**." This is published by Raspberry Pi itself, is well documented, and contains functions to access the different hardware modules contained in the RP2040 microcontroller. The installation procedure is complex, especially on Windows. Usage will be familiar to those who have programmed in Linux, but for others, there is a learning curve.

MICROPYTHON

This is the version of the Python programming language developed for microcontrollers. I appreciate its easy setup and programming, but it does not include support for all the hardware that we need (such as an LCD screen and FM radio modules). We could add third-party language modules to provide the missing capability, but managing them is complicated and is likely to change as new versions are released.

For these reasons, I opted not to use Python or **CircuitPython**, a version of the language developed by Adafruit.

ARDUINO IDE

The Arduino IDE is designed around the C++ language. Arduino microcontroller boards and the Arduino IDE have been popular among hobbyists for well over a decade, and as a result, many libraries have been published that allow an Arduino to communicate with different pieces of hardware.

The good news is that many of the Arduino IDE libraries also work on the Pico without any change. The bad news is that you must do some extra setup in the Arduino IDE to add Pico support. I decided that this setup was acceptable, so I have chosen the Arduino IDE. It has three advantages:

- Libraries are available for the hardware that we will use, with reasonably easy installation.
- Because it is based on C++, it's familiar to many people, and I expect it to remain available for many years.
- Pico-specific functions from the C/C++ SDK allow us to make full use of the Pico hardware.

Because we will be using the Arduino version of C, there is a good chance you may be able to run some of the sketches on an Arduino board in cases where we do not use any Pico-specific features or require a powerful microcontroller. However, I cannot guarantee that this will work.

SKETCHES AND LIBRARIES

We are using the word *sketches* for the code that people write on their microcontrollers, to distinguish them from **programs** running on your PC, such as the Arduino IDE.

A **library** contains blocks of microcontroller code. You can make use of them to perform tasks such as communicating with an LCD screen. Some libraries are available via the Arduino website, and some are written by individuals and made freely available elsewhere for everyone to use.

FINDING DOCUMENTATION

I don't have space in this book for an in-depth explanation of the Pico SDK, the C language, or the Arduino functions. I will only include documentation for our specific projects. If you want to know more, I suggest these sources:

The official Raspberry Pi Pico documentation can be found at www.raspberrypi.com/documentation/microcontrollers/.

The Arduino programming language is documented at www.arduino.cc/reference/.

You will find specifics of the Arduino system for the Pico at arduino-pico.readthedocs.io/.

For a reference guide to the C++ language, I suggest *C++ Pocket Reference*, by Kyle Loudon, published by O'Reilly and Associates.

INSTALLING THE IDE

Your first step is to install the IDE on your computer. Remember, the IDE is the development environment that enables you to edit and write sketches. You can think of it as being like a word processor, but for programming.

If you have used an Arduino microcontroller, you probably have a version of the Arduino IDE already installed. If it is version 2.0.0 or later, you should be able to use it for the Pico board as long as you follow my instructions in the section titled "Adding Code for Pico," below. However, my instructions have been tested with version 2.2.1. If you want to minimize the risk of any incompatibility, please use that version. If you are offered a later version on the Arduino website, you should be able to find that the site has archived previous versions, which you can download.

The IDE installation varies for different operating systems. I am providing instructions for Windows, Linux, and macOS. You can skip to the section that's relevant to you.

Do not plug in your Pico board yet.

WINDOWS INSTALLATION

Do not install the Arduino IDE through the Microsoft store, as that version is reportedly incompatible with the Pico.

1. Visit www.arduino.cc. At some point, Arduino may change the look and wording of its website, but I believe the basic options will still be the same.

2. If you are asked to accept cookies, make either choice that suits you.

3. On the menu bar, click **Software**. On this page, in the **Downloads** section, and beside **Arduino IDE 2.2.1** (or higher version number), you will see a **Download Options** section.

4. Select **Windows Win 10 and newer, 64 bits**. Note that in ***Windows 7***, I encountered a known bug that interfered with installation, and I don't know if it has been fixed; Windows 7 is no

longer supported by Arduino or Microsoft. All my tests on Windows computers were done with Windows 10.

5. After you select the version of your choice, you may be asked if you want to support Arduino IDE development with a donation. You don't have to. You can click **Just Download**.

6. Once the download finishes, you should have a file on your computer named **arduino-ide_2.2.1_Windows_64bit.exe** (or similar). This is the installer for Arduino IDE. Now, how do you find the file?

   Depending on your browser and your version of Windows, the download status bar may have an arrow mark allowing you to "Show in Folder" the file you just downloaded. Alternatively, there may be a download symbol in your menu bar, showing a down arrow and a line. Click this option if it appears; otherwise, look for the file in your Downloads folder or on the desktop, depending on your system settings.

   When you find the downloaded file, double-click it to open the installer, which will present a series of questions or directions:
   a) "License Agreement." You have to click **I Agree**.
   b) "Make the software available to all users or just yourself?" This is up to you.
   c) "Choose Install Location." The default is okay. Click **Install**.

   Installation now begins. This can take three or four minutes.

7. When the installation is complete, click the **Finish** button.

   You should be offered the option to launch the IDE. If not, look for the Arduino IDE shortcut, shown in Figure **4-5**, which was placed on your desktop if you allowed the installer to do so. Double-click the shortcut.

   If you cannot find the shortcut, launch the IDE from the list of programs in the Windows menu on your computer. It will be listed alphabetically as **Arduino IDE**.

4-5 *The shortcut for the IDE.*

If this is the first time you have installed the IDE, there are some additional steps you may need to complete:

8. Microsoft may ask you if you will allow it to make changes to your device. It is safe to agree. You may also receive a question from Windows Defender, asking if you wish to allow internet connections. Click the checkbox for **private networks**. Uncheck the box for **public networks**. Then click **Allow Access**. (You may be asked twice.)

9. You may be asked to allow software to be installed from Adafruit Industries. Click **Install**.

10. You may be asked to allow Arduino USB Driver to be installed. Click **Install**. (You may be asked twice.)

11. You may be asked to allow Genuino USB Driver to be installed. Click **Install**.

Now, the Arduino IDE should start, with its window showing a minimal sketch template, as shown in Figure **4-6**. Note that the black area at the bottom of the window will display status messages, including error messages. You can refer back to this figure for reference.

Windows users can skip the next two subsections describing Linux and Mac installations.

LINUX INSTALLATION

The easiest way to install Arduino on Linux is probably to use your Linux distribution's package manager. On Ubuntu or Debian, for example, you can do so by running the following command in a terminal:

```
sudo apt install arduino
```

If you are asked if you want your current account added to the **dialout** group, do so. This is required for permission to access serial ports.

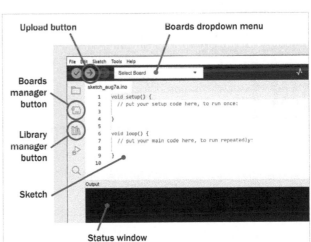

4-6 Opening screen for the Arduino IDE, with some features indicated that you are going to need.

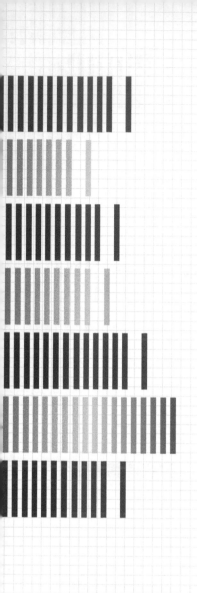

Currently, the above command installs the older version 1.8 of the Arduino IDE. I have successfully used this version with the Pico and the examples in this book. You can follow the instructions below, with the only issue being that the Library Manager and Boards Manager are launched using the menu at the top of the main window.

You can also follow Steps 1–3 from the Windows Installation instructions above and download the latest version of Arduino for Linux from there.

MacOS INSTALLATION

Visit www.arduino.cc and click the **Software** tab. Scroll to the **Downloads** section, and beside "Arduino IDE 2.2.1" (or higher version number), select either macOS Intel or macOS Apple Silicon, depending on your computer. This will download a file with the extension .dmg ("disk image"). Save it somewhere, for example on the desktop.

Double-click the file. This opens a window where you need to drag a file named Arduino IDE.app into the Applications folder to install it. When the installation finishes, you can close this window. Now you can find the Arduino IDE among your apps in the Launchpad. Double-click it to start it, and when asked if you want to run a program downloaded from the internet, click **Open**. When you get the prompt "Arduino IDE.app would like to access files in your Documents folder." click **OK**.

ADDING CODE FOR PICO

The remainder of our installation procedure should be the same for all computer operating systems.

Your next step is to install a library of code to enable the Pico to work with the LCD monitor that you will be using.

The IDE contains a feature named **Boards Manager** that will install the necessary code for you.

1. Leave the IDE window open, but go back to your browser window, where you will visit this GitHub repo, maintained by a programmer named Earle E. Philhower III:

    github.com/earlephilhower/arduino-pico

2. Scroll down this page to the **Installing via Arduino Boards Manager** section.

   Continue down to the subhead **Installation**.

   Now, you will see a long URL that looks like this: https://github.com/earlephilhower/ arduino-pico/releases/download/ global/package_rp2040_index.json.

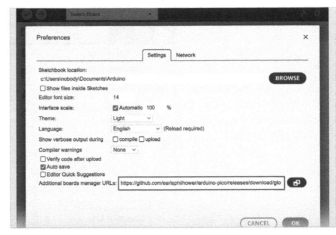

4-7 The Arduino IDE Preferences window.

   Don't type or click this URL. Instead, drag your mouse to highlight it, and press Ctrl-C (Command-C on a Mac) to copy it. Make sure you copy the whole thing.

3. Go back to the Arduino IDE window, which should still be open. Select the menu for **File > Preferences** (it may be named **Arduino IDE > Preferences** on a Mac), which opens a window like the one shown in Figure **4-7**.

4. Scroll down if necessary, and click in the Additional Boards Manager URLs field (outlined in red in Figure 4-7). If there is already text in it, move your cursor to the end of the text and type a comma.

5. Press Ctrl-V (Command-V on a Mac) to paste in the URL that you just copied, as shown in Figure 4-7.

6. Click **OK**. Nothing seems to happen at this point. You may not see confirmation that the URL you entered is valid, but this is normal.

7. In the Arduino IDE window, select **Tools > Board > Boards Manager** or click the **Boards Manager** button, as shown in Figure 4-6.

8. Wait a moment for the list to populate. If you have a slow internet connection, this may take a little while. Then, in the search bar at the top of the Boards Manager, type Raspberry Pi Pico.

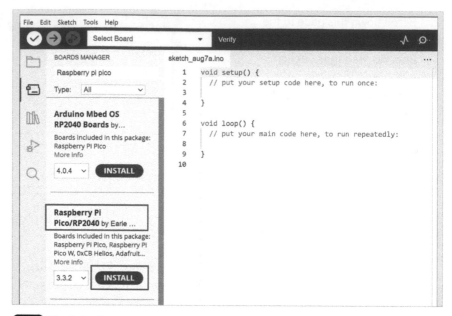

4-8 *Find the Raspberry Pi Pico code written by Earle F. Philhower III, and click the Install button.*

9. Now, you should see "Raspberry Pi Pico/RP2040, by Earle F. Philhower, III," outlined in red in Figure **4-8**. You may need to scroll down if it is preceded by other text.

    If you cannot find the right entry, check the small black status window at the bottom of the Arduino IDE for any error messages. If the long URL that you pasted in is invalid, you should get an error message there. Go back to your browser window, copy the URL again, and repeat the paste operation in the Boards Manager window.

10. **Install** the latest version of code for the Pico board by clicking the Install button. (You may need to scroll down further to find it.) There is a thermometer showing the progress. Note that if you have ever installed the code previously and then uninstalled the Arduino IDE, the code will not have been removed, and you won't need to reinstall it, although you may have an option to update it.

11. When the download is complete, close the Boards Manager by clicking the **Boards Manager** button at the left edge of the window.

12. Your Pico board should now be compatible with the Arduino IDE. To confirm that you can select it in the Arduino IDE, roll over this series of dropdown menus and click the final one:

```
Tools > Board > Raspberry Pi Pico/RP2040 > Raspberry
Pi Pico
```

If you are using a different version of the IDE, there may be fewer menus before you reach the one that includes "Raspberry Pi Pico." Although the list is otherwise alphabetical, you should find "Raspberry Pi Pico."

Be aware that in the Arduino IDE, the **Tools > Board** menu cleverly changes its content depending on the currently selected board. Just proceed to the last of the series of menus and select the Pico.

You have reached the end of this setup procedure. You should not have to repeat it.

YOUR FIRST PICO SKETCH

To test your Pico board, you can use a sketch named **Blink** that is included as a standard feature in the Arduino IDE. It simply blinks a tiny surface-mount LED located on the Pico board.

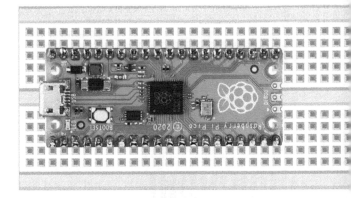

1. If you have not yet plugged the Pico board into your breadboard, you can do so now, as shown in Figure **4-9**. One way to do this without hurting your hand (or the board) is to put a wad of folded tissues over the Pico board before you press down firmly and evenly with your palm.

4-9 Place the Pico board on your breadboard exactly as shown.

2. Connect your Pico board with a USB port on your computer using the appropriate cable. Reminder: You must use a fully featured USB cable that can do data transfers, not just a phone-charging cable. When you bought your Pico, it may have been supplied with the correct type of cable. The small end, which is a Type-B Micro USB plug, fits into the metal socket at the left end of the board in Figure 4-9. Note that this type of plug only fits one way up; you may have to try it both ways.

3. Windows and Mac computers may show you a notification that a new disk drive has been connected. You can ignore this. If a new window has opened, close it now.

4. To load the **Blink** sketch into the Pico, select this menu option in the IDE:

   File > Examples > 01.Basics > Blink

   The sketch opens in a new window.

   Confirm that the Pico board is selected:

   Tools > Board: "Raspberry Pi Pico" > Raspberry Pi Pico/RP2040 > Raspberry Pi Pico

   If not, select **Raspberry Pi Pico**.

   Now you need to select a port on your computer to transmit data. For a brand-new Pico that has not been programmed before, the correct option will be **Tools > Port > UF2 Board**.

   A Pico that has previously been programmed from the Arduino IDE will show up as a serial port such as COM1, COM2, COM3, or COM4. If you see more than one serial port listed, choose the one with the highest number (which is likely to be COM4).

   If the IDE is recognizing your Pico board and is ready to communicate with it, you should see the USB symbol beside the name of the board, outlined in red in Figure **4-10**.

5. Click the **Upload** button in the Arduino IDE. This is the little arrow pointing to the right, highlighted in Figure 4-6.

   The upload process compiles the sketch into machine language, then uploads it to the microcontroller through the USB cable, and the sketch starts running automatically. You may have to be patient: The compile process can take as long as a minute.

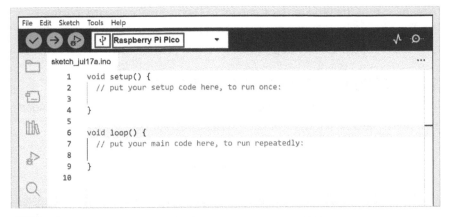

4-10 *When the USB symbol and the Pico board (outlined in red) are shown in the main IDE window, you can upload your sketch to the Pico.*

On a Mac, you may see a message:

`"Arduino IDE.app" would like to access files on a removable volume`.

If you get that message, click OK.

Now, you should see the onboard LED of the Pico blinking slowly. (Check Figure 4-4 if you have forgotten where the LED is located.) If this is all good, you can skip the rest of this section and proceed to the next one, telling you how to connect an LCD screen to the Pico.

TROUBLESHOOTING

You may encounter an error message in the IDE window telling you

`Upload Error: Failed Uploading: no upload port provided`.

When programming a new Pico microcontroller for the first time, it does not show up as a serial port to the computer. Select **Tools > Port > UF2 Board** and try again.

If you have programmed the Pico before, the **Tools > Port** menu should show you the available COM ports as described above. In some cases, you may need to reselect a COM port and then use the dropdown menu immediately below the menu bar to reselect the Raspberry Pi Pico.

Assuming that the little surface-mount LED on your Pico board starts flashing, your first sketch is working. You can close the Blink sketch window and skip troubleshooting steps 1–9.

TROUBLESHOOTING EARLIER VERSIONS OF THE IDE

If you are using an earlier IDE version such as 1.8, and the first-time upload of a sketch fails, you may have to do some troubleshooting, as follows:

1. If the Blink sketch is not still open, go back to the **File** menu and reopen it.

2. From the **File** menu, save the sketch to a new name, like **blink copy**, because example sketches cannot be modified under their supplied names.

3. Go to the menu option **Sketch > Export Compiled Binary**. This compiles the sketch and saves it.

4. Go to the menu **Sketch > Show Sketch Folder**. This opens a window where you should see the blink copy sketch that you just saved. If there is a folder named **build**, open it and then open the folder contained within that starts with **rp2040**. Find a file with the extension **.uf2**. (Windows may hide file extensions by default.) This file contains your sketch in a format suitable for uploading to the Pico.

5. In Windows 10, click the USB icon at the corner of your taskbar. (The icon looks like a little USB plug. It may be hidden until you click the arrow mark.)

6. You should see the option **Open Devices and Printers**. Click that option, and in the window that opens, look for an external drive listed under a name such as **RP1** or **RP2**. (On a Mac, you instead open the Finder and locate the removable drive RPI-RP2 in the left part of the window.)

7. Click that drive to open a window for it.

8. If you don't find it, unplug the USB cable to the Pico. Press and hold down the **BOOTSEL** button on the Pico board (see Figure 4-4) while you plug the USB cable back in again. Release the **BOOTSEL** button when the USB is connected. Now, a window for the Pico microcontroller should open.

9. Drag and drop your **blink copy** file into the Pico window. This should upload the sketch to the Pico microcontroller, which should run the sketch, blinking the LED.

   Now that the IDE has figured out how to find the Pico board, the Pico should show up in the IDE's **Tools > Port** menu and in the dropdown menu at the top of the IDE window. You can select it there. From this point onward, the Arduino IDE should be able to upload sketches directly to the Pico by default. You can try it by clicking the **Upload** arrow button. The sketch should compile and upload into the microcontroller automatically, overwriting the previous copy, and the LED should continue blinking.

PICO BOARD FEATURES

To get to know the Pico and to prepare for the following projects, I want you to connect the Pico to an *LCD screen*. LCD stands for "*liquid crystal display*."

I will refer to the solder pads around the edge of the Pico in the closeup photograph of the board shown in Figure **4-11**, as *pins* because they have pins soldered into them, connecting with your breadboard. The pins are identified by number, but also by abbreviations that refer to their function. I will explain more of the abbreviations as we go along, but for now, you need to know what these ones mean:

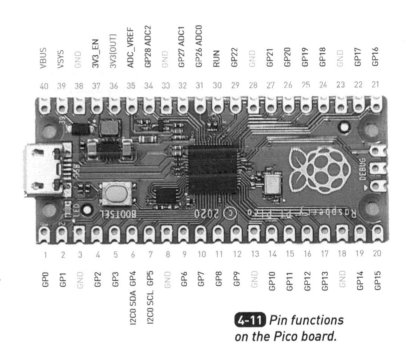

4-11 *Pin functions on the Pico board.*

VSYS is the system voltage, which is 5VDC. The Pico board takes 5VDC through the USB cable from your computer and routes some of it to this pin so that you can power external devices from it. *Do not attach a battery or AC adapter to the Pico board*. Your computer supplies all the power it needs.

GND is negative ground, also shared with your computer. The Pico board has multiple GND pins for your convenience. All these pins are connected with each other internally on the board.

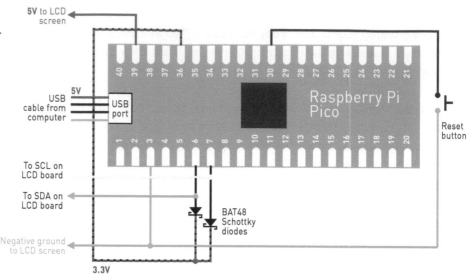

4-12 *Schematic for connections from Pico board to LCD.*

5V to LCD screen

USB cable from computer

5V

USB port

Raspberry Pi Pico

Reset button

To SCL on LCD board

To SDA on LCD board

BAT48 Schottky diodes

Negative ground to LCD screen

3.3V

Pins 6 and 7, labeled **I2C**, enable the board to communicate with external devices such as your LCD. **I2C** is the name of a communications protocol that is commonly used by microcontrollers.

Pins identified with **GP** are general-purpose pins, which are configured by commands in a sketch that has been uploaded into the microcontroller.

Pin 36, labeled **3V3OUT**, provides a 3.3V supply to any external component that needs it. The Pico microcontroller is actually a 3.3V device; there is a voltage regulator built into the board that converts 5V from your USB connection to 3.3V for the microcontroller.

(Note that the microcontroller can be overloaded and damaged permanently if you accidentally supply 5V to any of the pins around the board. Be careful when making connections.)

Because I am using breadboards turned horizontally throughout this book and refer to their top, bottom, and left edges while the board is horizontally oriented, I encourage you to orient your breadboard the same way.

Your Pico board should be plugged in so that it is two columns of holes from the left edge of the breadboard, as shown in Figure 4-9.

HOOKING UP YOUR LCD SCREEN

Figure **4-12** shows a schematic for a simple circuit using jumpers and wires to connect the Pico board with the little board on the back of the

LCD screen. The same circuit is shown pictorially in Figure **4-13**.

Unplug your USB connection before you start building this circuit.

Your first step is to make the four connections between your breadboard and the mini-board on the back of the LCD. I strongly recommend color-coding these connections. You can use male-to-female jumper wires for this purpose. Each of these wires has a single-pin plug at one end, which you insert into the breadboard, and a socket at the other end, which you slide onto one of the four connection pins on the back of the LCD.

The red wire supplies 5V power from the Pico board to the LCD, and the blue wire shares negative ground. The green and yellow wires will send data to the LCD using the I2C protocol. The three-letter abbreviations in Figure 4-13 will match the text beside pins in the LCD board, as you can see in Figure **4-14**.

Looking back at the schematic in Figure 4-12 and the breadboard layout in Figure 4-13, you will see two diodes between Pins 6 and 7 and the red bus on the breadboard. These diodes are included to protect the Pico board from any excessive voltage that may return from the LCD. Remember, the LCD is a 5V device, but inputs on the Pico board must never be connected with a voltage greater than 3.3V.

The positive bus on the lower edge of the breadboard is supplied with 3.3V from Pin 36 of the Pico board. The 3.3V connections are shown as dashed lines to remind you that they are 3.3V, not 5V.

(Note that in In our schematics, solid red wires have 5V, while dashed red lines have 3.3V.)

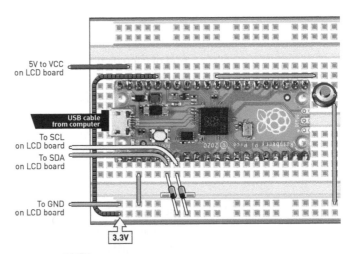

4-13 Breadboard connections from Pico board to LCD.

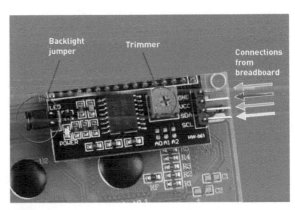

4-14 The miniboard on the rear of the LCD.

If the green or yellow wire from the LCD exceeds 3.3V, the diode shunts that excessive voltage to the positive bus. It may seem confusing that the negative end of each diode is connected to the positive bus, but remember, we only want the diode to conduct current if the green or yellow wire carries more than 3.3V. Under that condition, the 3.3V positive bus will be "more negative."

We use Schottky diodes for this purpose because they have a lower threshold voltage than diodes such as the 1N4148, which are used in logic circuits.

A pushbutton (properly known as a **momentary switch**) is connected between Pin 30 and the negative bus at the bottom of the board. When you hold down the pushbutton, this grounds Pin 30 and resets the Pico. There is a **pullup resistor** inside the Pico board, so you don't need to add one yourself for the pushbutton.

SCREEN TEST

Now, finally, you are almost ready to send text to the LCD. There is just one more step: Install the code that will enable the Pico board to communicate with the screen. (You can think of it as being like a printer driver on a PC.) To obtain this code, first you must find it online and download it onto your computer.

1. In your IDE window, choose **Sketch > Include Library > Manage Libraries...** or click the Library Manager button to open the library installer side panel (or window, in IDE version 1.8 or older).

2. In the search field under the Library Manager heading, type

    hd44780

    as a search term. The hit that you want will not be the first one in the list. You may have to scroll down till you find

    hd44780 by Bill Perry

    This is outlined in red in Figure **4-15**.

3. Click **Install**.

4. When the installation is complete, close the library installer side panel by clicking the Library Manager button again.

5. The library has included a test program named **Hello World**. You can find it now in a menu on the IDE. Select

   File > Examples > hd44780 > ioClass > hd44780_I2Cexp > HelloWorld

   (In the Examples menu, **hd44780** may be at the bottom of the list, even though the rest of the list is alphabetical.) The sketch now appears in a new IDE window.

6. Go back to

   Tools > Board > Raspberry Pi Pico/RP2040 > Raspberry Pi Pico

   in the main IDE window, and make sure that the Pico is still selected.

7. In the Hello World sketch window, click the Upload arrow to send the **Hello World** sketch to the microcontroller. The sketch will start to run, but you may not see anything on the screen yet.

8. On the back of the LCD is a trimmer that controls the brightness of the screen. (The trimmer is circled in Figure 4-14.) You will need a very small flat-blade screwdriver to turn it. Insert the screwdriver, and hold it in place while you turn the screen over so that you can see the front of it. Rotate the screwdriver one way or the other, and you will find the sweet spot where the screen is not too bright or too dark, and you see the text Hello, World!

4-15 You need to install the code outlined in red to drive your LCD display.

```
1  // Wire.h is a library for I2C and is part of Arduino
2  #include <Wire.h>
3  #include <hd44780.h> // the general LCD library
4  #include <hd44780ioClass/hd44780_I2Cexp.h> // LCD library for I2C-equipped LCD
   modules
5
6  hd44780_I2Cexp lcd;
7
8  // define the size of the display
9  const int LCD_COLS = 16;
10 const int LCD_ROWS = 2;
11
12 void setup()
13 {
14   int status;
15   status = lcd.begin(LCD_COLS, LCD_ROWS);
16   if(status) // non zero status means it was unsuccessful
17   {
18     hd44780::fatalError(status); // does not return
19   }
20
21   // Print a message to the LCD
22   lcd.print("First row.");
23   // Go to the second row
24   lcd.setCursor(0, 1);
25   // Print another message
26   lcd.print("Second row.");
27 }
28
29 void loop()
30 {
31 }
```

4-16 *Sample code for using the LCD display with the Pico.*

IF IT DOESN'T WORK

In Figure 4-14, I have circled a little black tab connecting two pins in the board on the back of the LCD. This jumper must be present to activate the backlight on the LCD screen. If the jumper is missing, the screen won't be illuminated. You have to short the pins together. You may be able to do this with an alligator clip or by carefully wrapping a piece of wire around the two pins and using sharp-nosed pliers to compress it and hold it in place.

If the display is not connected properly, perhaps because of a wiring error, the LCD library will notice that the I2C connection failed and will report an error by blinking the Pico's onboard LED four times in sequence, then a longer pause, after which the blinking sequence repeats.

UNDERSTANDING THE SKETCH

The **Hello World** sketch contains comments explaining the statements in the sketch. More extensive documentation of the library can be found from inside the Arduino IDE, under **File > Examples > hd44780 > Documentation**.

A sample sketch similar to the **Hello World** example you just ran is shown in Figure **4-16**. I have included a short summary of its functions below. If you are not very interested in how the I2C connection works, because you are impatient to start using the Pico for radio projects, feel free to skip the rest of this chapter. You can always come back to it later.

The **Wire** library, included on line 2 of the sketch, handles I2C communication. By default, it uses Pico Pins 6 and 7 for the I2C clock and data lines. The following two lines include the display library. On line 15, the function call

```
status = lcd.begin(LCD_COLS, LCD_ROWS);
```

initializes the display and tells the library what size the display is. Our LCD screen is a "1602" display, which means it has 16 columns and 2 rows. The function **lcd.begin** returns 0 if the display was successfully initialized, and this is checked on line 16. The **lcd.setCursor(column, row);** function call on line 24 can move the cursor to any position on the display. The row and column numbers start at **0,0** in the top left corner. The cursor is not visible (unless turned on by a command to the display) but marks the place where any future print command will place its text. **lcd.print()** is used to send text or display variable values to the screen. When sending text to the LCD, any characters already on the screen in the character positions beyond the sent text are left unchanged, which may have confusing effects. One can send strings padded with spaces, or use **lcd.clear();** to clear the whole display.

The LCD display we're using here contains a standard LCD controller chip called HD44780. The display itself consists of a grid of pixels, each of which may be on or off (dark or light). The controller chip sends signals to the display to turn the pixels on or off, rapidly scanning the pixel rows. The controller has an internal RAM memory for the characters that should be displayed and a read-only memory containing character tables—the pixel patterns corresponding to each character. The controller is said to be character based, because you can send commands like "Place the character **b** at the current cursor position" without having to specify which pixels are needed to make a **b**. It is also possible to define up to eight custom characters.

The HD44780 controller on its own has a parallel interface, meaning data is sent to it using either four or eight data lines and some control lines. The HD44780 datasheet explains in detail how to send text to the display and position it. It also explains which characters are available and some special effects: inverted text, blinking text, and a blinking cursor.

To save wiring work and microcontroller IO pins, we use a display with an I2C adapter. The adapter is widely available on eBay and through Asian sources and is often sold together with LCD modules. Buy an LCD module with the I2C interface already soldered on.

The LCD connected with the breadboarded circuit is shown in Figure **4-17**.

STATUS UPDATE

This experiment was almost entirely a process of setup to provide you with some capabilities that you will need in the rest of the book. You should have picked up some knowledge about using a microcontroller (if you did not have that information already), and you saw how it can display messages on a screen. The steps to get the Pico ready to run may be useful if you decide to use it for purposes outside of radio.

Up next: using the Pico as the basis for a radio transmitter.

4-17 *Successful completion of the LCD test program.*

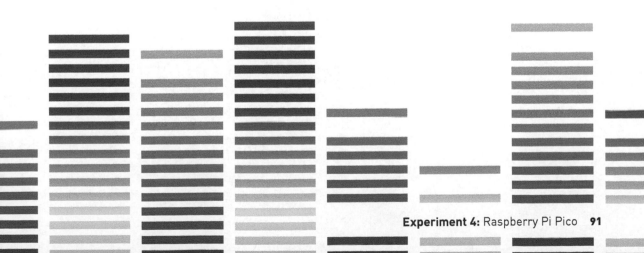

5

A PICO TEST
TRANSMITTER

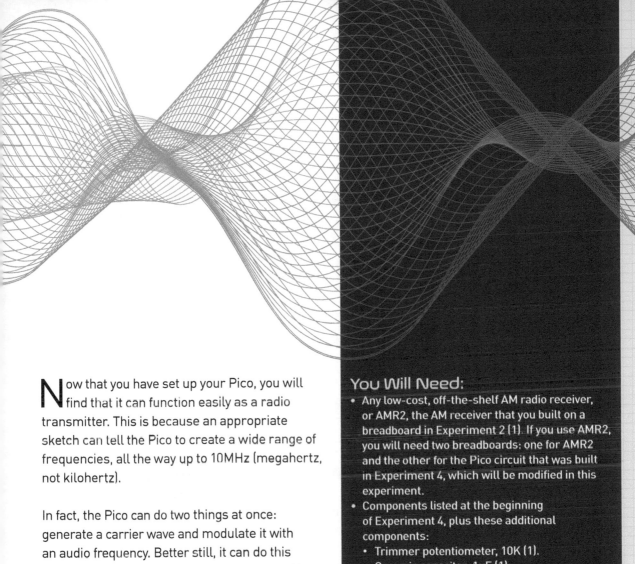

Now that you have set up your Pico, you will find that it can function easily as a radio transmitter. This is because an appropriate sketch can tell the Pico to create a wide range of frequencies, all the way up to 10MHz (megahertz, not kilohertz).

In fact, the Pico can do two things at once: generate a carrier wave and modulate it with an audio frequency. Better still, it can do this precisely, unlike the 555 timers that you used in Experiment 1, which are impossible to control accurately.

The microcontroller only needs about 40 lines of code to tell it how to perform these tasks. After you install that code, I'll take you through it, line by line, to explain how it works. You will be learning the basics of programming so that you can modify a sketch or write one of your own. You will also see how a feature built into the Pico, known as pulse-width modulation, can be adapted as a way to deliver test tones via a radio transmission.

You Will Need:

- Any low-cost, off-the-shelf AM radio receiver, or AMR2, the AM receiver that you built on a breadboard in Experiment 2 (1). If you use AMR2, you will need two breadboards: one for AMR2 and the other for the Pico circuit that was built in Experiment 4, which will be modified in this experiment.
- Components listed at the beginning of Experiment 4, plus these additional components:
  - Trimmer potentiometer, 10K (1).
  - Ceramic capacitor, 1μF (1).
  - Resistor, 100 ohms (1).

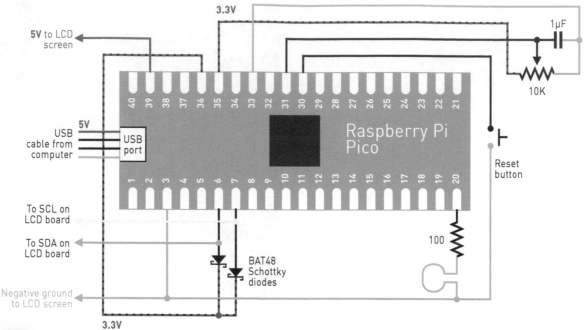

5V to LCD
screen

USB
cable from
computer

5V

To SCL on
LCD board

To SDA on
LCD board

Negative ground
to LCD screen

3.3V

3.3V

1μF

10K

Raspberry Pi
Pico

Reset
button

100

BAT48
Schottky
diodes

5-1 *This circuit is built by adding components to the circuit from Figure 4-12.*

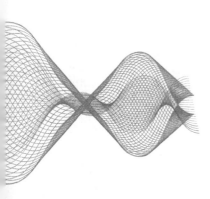

SIGNAL TEST

In Figures **5-1** and **5-2**, three components have been added to the circuit from Figures 4-12 and 4-13. This is all you need to convert your Pico into a transmitter because a new sketch that you will upload to the Pico will do the rest.

The trimmer potentiometer will adjust the transmission frequency, and the loop of blue wire (about 6") at the bottom of the board is the antenna, transmitting the signal to any nearby AM receiver. Obtain the sketch "5-PWM_frequency_generator"shown in Figure **5-3** from my GitHub page at github.com/fjansson/MakeRadio, and upload it to your Pico. It will automatically overwrite (replace) any sketch that is already in the Pico memory.

As soon as the upload is complete, the Pico will execute the instructions in your sketch, and you should see a number on the LCD screen that tells you the carrier frequency in kHz. Turn the trimmer, and the frequency should vary from 500kHz up to 2,000kHz, covering the AM broadcast band, with some extra margins.

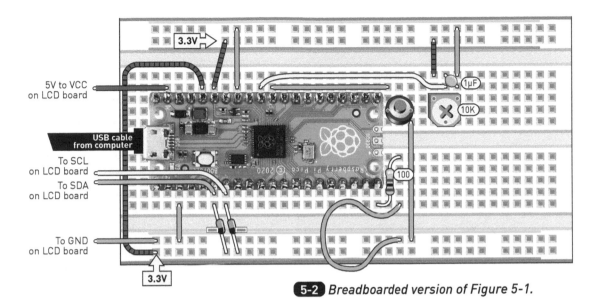

5-2 *Breadboarded version of Figure 5-1.*

Labels on figure:
- 3.3V
- 5V to VCC on LCD board
- USB cable from computer
- To SCL on LCD board
- To SDA on LCD board
- To GND on LCD board
- 3.3V
- 1μF
- 10K
- 100

Set the trimmer to the middle of its range. Now, you can pick up the signal with any portable radio or the AMR2 circuit that you built previously. (Be sure to disconnect the 555-chip test signal generator in that circuit to avoid confusion.)

The range of your transmission should be about 3 feet. When you tune in to it, you'll hear a series of rapid beeps. The Pico is doing three things:
- Generating the carrier frequency.
- Switching the carrier on and off rapidly to create an audio tone.
- Adding tiny pauses in the tone to create the beeps.

If you are using AMR2, you can now find out what its range of frequencies really is. Turn the tuning capacitor of AMR2 fully counterclockwise, and then turn the trimmer potentiometer on your Pico transmitter until the beeps are as loud as possible. AMR2 and your test transmitter are now tuned to the same frequency, which is shown as the number on your LCD screen. Make a note of it.

```
1  #include <Wire.h>
2  #include <hd44780.h>
3  #include <hd44780ioClass/hd44780_I2Cexp.h>
4
5  hd44780_I2Cexp lcd;
6  const int LCD_COLS = 16;
7  const int LCD_ROWS = 2;
8
9  const int output_pin = 15;
10
11 void setup()
12 {
13   int status;
14   status = lcd.begin(LCD_COLS, LCD_ROWS);
15   if(status) // non-zero status means it was unsuccessful
16   {
17     hd44780::fatalError(status);
18   }
19 }
20
21 void loop()
22 {
23   float trimmer = analogRead(A0)/1023.0;
24   int frequency = trimmer*1500000 + 500000;
25
26   lcd.clear();
27   lcd.print(frequency/1000);
28   lcd.print(" kHz");
29
30   analogWriteFreq(frequency);
31   analogWriteRange(4);
32
33   int i;
34   for (i = 0; i < 100; i++) // 200 ms tone
35   {
36     analogWrite(output_pin, 0); // duty cycle  0% - output no carrier
37     delay(1);
38     analogWrite(output_pin, 2); // duty cycle 50% - output carrier wave
39     delay(1);
40   }
41
42   delay(100); // 100 ms silence
43 }
```

5-3 *Sketch to make the Pico into a radio transmitter.*

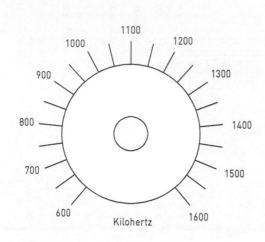

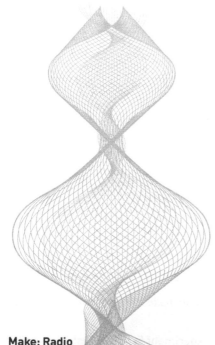

5-4 *A tuning dial for your AMR2 radio. Your actual range of frequencies, and the intervals at which they are spread around the dial, will depend on your particular components.*

If you have difficulty finding the trimmer position that creates the loudest beeps, try moving the receiver away from the transmitter.

Repeat the procedure with the tuning capacitor wheel turned fully clockwise on AMR2, and readjust your trimmer potentiometer until the signal is as loud as possible. Make a note of the new number on your LCD screen. Now, you can calibrate AMR2. Adjust the trimmer on your Pico circuit in steps of 50kHz, as shown on the LCD screen. Adjust the trimmer capacitor on AMR2 to match, and you should be able to make a tuning dial for AMR2 similar to the one shown in Figure **5-4**.

UNDERSTANDING THE SKETCH

Lines 1, 2, and 3 of the sketch declare that it's going to include the **library** for your LCD screen. This library should still be in the Arduino IDE, where you downloaded it in Experiment 4. *You must complete Experiment 4 to make all the other sketches in this book recognize the LCD screen. Otherwise, the sketches won't work.*

Lines 6 and 7 create two **constants** named LCD_COLS and LCD_ROWS as **integers**, giving them values of 16 and 2, respectively. This defines the number of columns and rows on your screen. If you use a different screen later, you could change these values.

• A constant keeps its value while the program is running.
• An integer is a whole number without a decimal fraction.

Line 9 defines another constant, selecting the output pin that we chose to connect with your antenna. We are using GP15. (See Figure 4-11.)

(Notice that the pin numbers in a sketch are not the same as the basic pin numbers on the Pico board.)

Continuing down the listing, all Arduino sketches contain two **functions** named **setup** and **loop**.

(A function is a routine that is contained within two curly brackets, properly known as **brace** symbols.)

Additional functions can be added and called by name, but you only need **setup** and **loop** in this sketch.

The setup prepares everything, and the loop then repeats indefinitely. In this sketch, the setup just initializes the LCD screen. Then the loop begins, and repeats between the braces on line 22 and line 43.

On line 21, the term **void** simply means that a function is beginning that doesn't return a result.

Line 23 defines a **variable** named **trimmer**, which will represent the position of your trimmer potentiometer.

- A variable is like a constant—a value stored in memory—except that the sketch enables its value to change.

On line 23, **analogRead** is the name of a function that is built into the Pico. It reads the voltage of the analog input A0, which is wired to the trimmer. The Pico is a digital device, but it can look at one of its analog input pins and see what the voltage is. This voltage can range from 0V to 3.3V.

In your circuit, analog input A0 is wired to the **wiper** of your trimmer, and because the trimmer is placed between the 3.3V positive bus and the 0V ground bus, you can be sure that the input will indeed range between 0V and 3.3V.

The **analogRead** function not only reads the voltage but also automatically converts it into an integer between 0 and 1,023. This is an **analog-digital conversion**. On line 23, the integer is divided by 1,023.0 to create a decimal fraction between 0 and 1.

For example, suppose the voltage is 1.1V (if the trimmer is turned to one-third of its range). Then the digital value will be 341. Divide by 1,023.0, and you get 0.333. This value is assigned to the trimmer variable. The fraction makes it easier to do the next calculation.

On line 24, the sketch creates an integer variable named **frequency** by multiplying the trimmer variable by 1.5 million and adding 500,000. This is the range you need to generate a carrier wave.

Lines 26 through 28 display the carrier frequency on your LCD screen in kHz (after dividing the frequency by 1,000).

On line 30 and 31, **analogWriteFreq** and **analogWriteRange** are built-in functions that set up the frequency and resolution to be used by future calls to **analogWrite**. The sketch is getting everything ready to build a carrier wave.

Line 33 sets up a new integer variable named **i** that will be used to count the number of cycles of a new loop between lines 35 and 40. Variable **i** will begin with value 0 and count up to 99 before the Pico exits the loop at line 40.

What happens in this loop? Remember that **output_pin** is a constant with a value defined on line 9 of the sketch. So **analogWrite** will send a value of 0 through the output pin, which turns it off. Then *delay(1)* adds a pause of 1 millisecond ($\frac{1}{1000}$ of a second). Then, on line 38, **analogWrite** sends a value of 2 through the output pin, followed by another delay of a millisecond. This procedure is repeated 100 times. Its purpose is to switch the carrier wave off and on 100 times at a frequency of 500Hz, as shown (approximately) in Figure **5-5**. The effect of this is to create an audio tone in a radio receiver tuned to the right carrier frequency.

But how was the carrier wave frequency set? By the value assigned to the **frequency** variable on line 24. I'll explain more about how this happens in the next section, below, on pulse-width modulation.

After the loop repeats 100 times, line 42 inserts a pause lasting 100 milliseconds—that is, $\frac{1}{10}$ of a second. This creates a gap between one beep and the next.

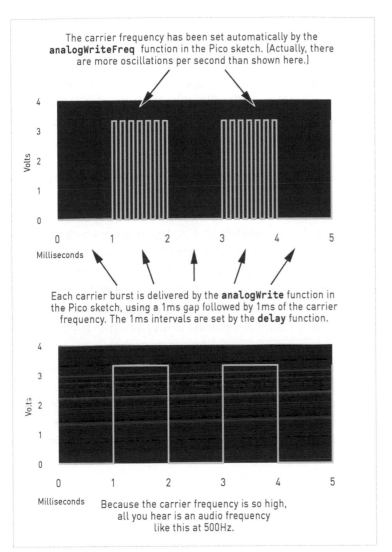

The carrier frequency has been set automatically by the **analogWriteFreq** function in the Pico sketch. (Actually, there are more oscillations per second than shown here.)

Each carrier burst is delivered by the **analogWrite** function in the Pico sketch, using a 1ms gap followed by 1ms of the carrier frequency. The 1ms intervals are set by the **delay** function.

Because the carrier frequency is so high, all you hear is an audio frequency like this at 500Hz.

5-5 *Lines 36–39 of the sketch create an output from the Pico like this.*

At line 43, the main loop ends—but because it began with the special name **loop** on line 21, it repeats itself automatically.

The principle of this sketch is similar to the principle of the circuit that you built with a pair of 555 timers in Experiment 1. In that circuit, a fast timer generated the carrier wave, while a slower timer switched it on and off at an audio frequency. In your sketch, lines 36–39 do something similar.

How? By using a feature called pulse-width modulation, which is built into the Pico.

PULSE-WIDTH MODULATION

If you have a stream of pulses looking like a square wave, you may want to vary the length of each "on" pulse while keeping the frequency constant. This could be useful, for instance, to adjust the apparent brightness of an LED. If the pulses are fast enough, the LED will look as if it is glowing constantly, but if the pulses are only "on" for half the time, the LED will be half as bright. This would be a *50% duty cycle*.

The duty cycle of a square wave is the percentage of the time that the output is high.

If you can reduce the pulse width (while keeping the frequency constant) so that the pulses are only on for one-quarter of the time, you have a 25% duty cycle, and so on. This is a very efficient way to dim an LED. The technique is known as *pulse-width modulation*, usually abbreviated as **PWM**, and it can also be used to control some types of motors. In fact it is such a common task for microcontrollers, many of them include it as a built-in feature.

Our sketch for the radio test circuit uses the PWM feature of the Pico as a convenient way to create a square wave output. This is delivered with the **analogWrite()** function, which has *analog* in its name because sometimes you may want to adjust the voltage of the output to a level somewhere between 0V and 3.3V, although we don't need that feature in this application.

However, we did need to set the duty cycle.

First, the function **analogWriteRange(4)** assigned the arbitrary number 4 on line 31.

Then, **analogWrite(output_pin, 0);** on line 36 set a duty cycle of 0% (switching the carrier wave off), followed by a delay of 1 millisecond, and then, **analogWrite(output_pin, 2);** on line 38 set the duty cycle of the carrier wave to 50% because 2 is 50% of 4 and the range was set to 4 initially. Then there was another 1-millisecond delay, keeping the carrier wave on for that period. Because the bursts of the carrier wave are so rapid, they create an audio frequency.

Finally, line 42 interrupted the audio frequency by leaving the carrier switched on for 100 milliseconds (1/10 of a second) without any interruptions. Since the carrier amplitude is constant, without any gaps in it, you hear

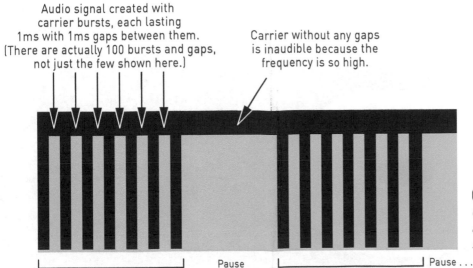

Audio signal created with carrier bursts, each lasting 1ms with 1ms gaps between them. (There are actually 100 bursts and gaps, not just the few shown here.)

Carrier without any gaps is inaudible because the frequency is so high.

5-6 *Bursts of carrier frequency are now shown as solid rectangles.*

Beep for 200ms

Pause for 100ms

Beep for 200ms

Pause . . .

nothing during this interval. I like to keep the carrier on during these breaks because it prevents the radio receiver from picking up other stations.

This is illustrated in Figure **5-6**, where the scale has been reduced compared with Figure 5-5, so the carrier bursts are now displayed as solid blocks. The result is, you hear a series of little beeps instead of one continuous tone.

MORE INFORMATION

For more information on how **analogWrite** works on the Pico, look up "Analog Outputs" in the Arduino-Pico documentation (arduino-pico .readthedocs.io). To find the details of the PWM hardware—for example, which pins can produce independent PWM signals—see the Raspberry Pi Pico datasheet.

STATUS UPDATE

In this experiment, you learned how to use the Pico to generate square waves, and you built an AM test transmitter with frequency readout on an LCD screen, creating the possibility of calibrating the AMR2 receiver. But maybe it would be nice to transmit something more than beeps. I promised that the Pico could add audio to a carrier wave, and in the next experiment, you will see how. And if you remember that I mentioned the need for a carrier to be a pure sine wave, you'll find we can take care of that, too.

6

AUDIO
TRANSMISSION
VIA PICO

In Experiment 3, you built AMT1, which modulated a carrier wave with an audio frequency, and broadcast it in the AM broadcast band. However, its carrier was not very stable, and you had no way of knowing exactly what its frequency was.

In Experiment 5, you used a Raspberry Pi Pico's pulse-width modulation (PWM) hardware as a better way to generate a carrier wave, enabling you to set the frequency accurately. You then modulated it to create a series of beeps.

The next step is to transmit speech or music via the Pico. It can't synthesize these complicated signals on its own, but it can receive an analog audio input from a source such as a music player and digitize it. After the signal has been digitized, it can create a modulated carrier that is then filtered to form a fairly good imitation of a sine wave. Finally, you will have a "real" radio transmitter, which I am calling AMT2B.

You Will Need:

- The circuit from Experiment 5, which I'll refer to now as AMT2A, with Pico and LCD screen on a breadboard (1).
- An AM radio receiver, or AMR2 from Experiment 2 (1).
- Ceramic capacitors: 220pF (1), 47nF (1), 1µF (1).
- Electrolytic capacitor, 10µF (1).
- Resistors: 1K (3), 10K (2).
- Tuning capacitor, 200pF, type 223P (1).
- 2N3904 bipolar NPN transistor (1).
- Nonconductive and nonmagnetic cylinder about 3.5" in diameter, such as a plastic vitamin bottle, for winding the coils (1).
- 22- or 24- gauge hookup wire for coil (30 feet).
- 1/8" audio jack with screw connection, or breadboard audio jack (1).
- Audio cable with male 1/8" plug at each end, mono or stereo (1).
- Source for music or spoken word, such as a music player, a phone with a 1/8" audio jack, or the laptop that programmed the Pico.

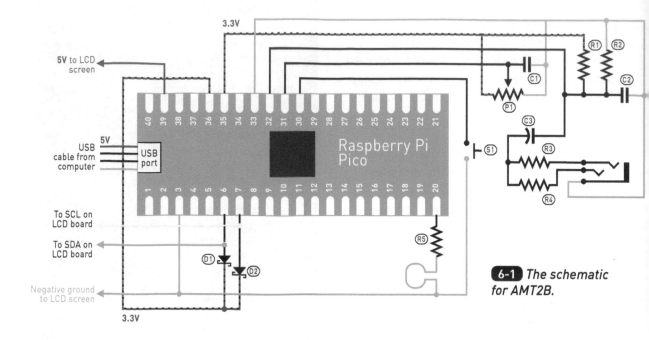

6-1 *The schematic for AMT2B.*

CONSTRUCTION

The schematic of AMT2B is shown in Figure **6-1**, and the breadboarded version is shown in Figure **6-2**.

The circuit is the same as for the tone transmitter in the previous experiment, with the addition of four resistors, two capacitors, and a schematic symbol representing an audio jack socket. This should be the type that has screw terminals, which you used in Experiment 3. (Two versions were shown in Figure 3-4.) Be careful to ground the appropriate terminal of the jack socket. The other two terminals can be used either way around, as they are connected together through R3 and R4.

After checking your wiring carefully, connect the Pico to a computer with a USB cable and upload the 6-AM_PWM" program in Figure **6-3**, which you can copy from my GitHub page at github.com/fjansson/MakeRadio.

To test the transmitter, you need a source of speech or music. If you program the Pico with a laptop, the computer probably has an audio output you can use for this purpose. The same warnings as in Experiment 3 apply, but I don't think there are any hazards associated with your USB port.

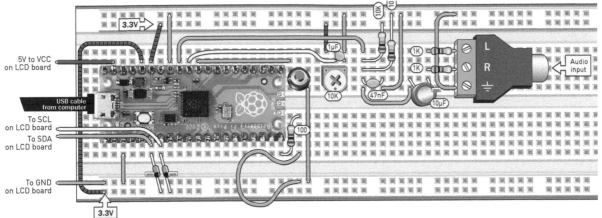

6-2 *Breadboard view of AMT2B.*

Play some music with your audio source, and turn up the volume. Place an AM receiver such as AMR2 near the transmitter. AMT2B shows the transmitter frequency on the LCD screen, and you can adjust it using the trimmer.

Try to find the transmitted signal on your receiver. If you hear other stations on the same frequency, adjust the transmitter and receiver until you find a quieter frequency.

The LCD screen now shows you three values. On the top row is the carrier frequency in kilohertz, which you can adjust using the trimmer. On the bottom row is a value labeled **div**, which we'll return to in a moment, and value, labeled **peak**. The peak value is a measure of the peak audio amplitude as a percentage of the full-scale value. Adjust the volume of the audio source so that the peak value is at about 90% (or, if your audio makes this difficult, just try to avoid reaching 100%). The audio quality is best when the signal utilizes the available range without exceeding

```
1  #include <Wire.h>
2  #include <hd44780.h>
3  #include <hd44780ioClass/hd44780_I2Cexp.h>
4  #include <hardware/clocks.h>
5  hd44780_I2Cexp lcd;
6  const int LCD_COLS = 16; // LCD geometry
7  const int LCD_ROWS = 2;
8  const int outPin = 15;   // the output pin we want
9  const float f_min = 530000;  // low f limit   530 kHz
10 const float f_max = 1700000; // high f limit 1700 kHz
11 const float scale = 2.0;     // volume control
12 float f_sys;     // system clock frequency
13 float peak = 0;  // measured peak audio value
14 void setup() {
15   int status = lcd.begin(LCD_COLS, LCD_ROWS);
16   if(status)
17     hd44780::fatalError(status);
18   gpio_set_drive_strength(outPin, GPIO_DRIVE_STRENGTH_12MA);
19   f_sys = clock_get_hz(clk_sys); // system clock in Hz
20 }
21 void loop() {
22   char str[17];
23   float trimmer = analogRead(A0)/1023.0;
24   float f = f_min + (f_max-f_min) * trimmer; // in Hz
25   int divisor = f_sys / f;
26   f = f_sys / divisor;  // calculate frequency back
27   analogWriteFreq(f);        // set frequency
28   analogWriteRange(divisor); // set analog range
29   lcd.clear();          // clear, cursor to top left
30   snprintf(str, 17, "%6.1f kHz", f/1000);
31   lcd.print(str);
32   lcd.setCursor(0, 1);  // place cursor on second row
33   snprintf(str, 17, "div%4d peak%3.0f%%", divisor, peak*200);
34   lcd.print(str);
35   peak = 0;  // reset peak value
36   for (int i = 0; i < 20000; i++)
37   {
38     float in = analogRead(A1) / 1024.0;
39     in = (in - 0.5) * scale + 0.5; // scale the input value
40     peak = max(peak, abs(in-0.5)); // keep track of the peak
41     in = constrain(in, 0, 1);      // constrain to range 0..1
42 //  int out = in * 0.5 + divisor;  // duty cycle - simple
43     int out = asinf(in) / M_PI * divisor; // correction
44     analogWrite(outPin, out);
45     //delayMicroseconds(5);
46   }
47 }
```

6-3 *Sketch for AMT2B, the Pico AM transmitter.*

it. If the audio signal exceeds the capability of your circuit, the highest or lowest parts will be clipped or cut off, degrading the audio quality.

You may remember that clipping was a potential problem for AMT1. When you turned up the volume too high, it caused distortion, but there was no reliable way of measuring or predicting it. By displaying the peak value on the screen, the Pico transmitter gives a way of optimizing the volume.

When the transmitter and the receiver are tuned to the same frequency, and the audio signal range has been adjusted, move the transmitter and receiver apart to see the range of the transmission. At this stage, it might be around 3 feet. You can increase the range in two steps: by improving the antenna, and by adding a transistor amplifier. Before getting there, I'll explain how the transmitter works, starting with how the audio signal enters the Pico.

HOW IT WORKS

As in AMT1, the audio signal is supplied to the transmitter through an audio jack, and the left and right audio channels from the audio jack pass through 1K resistors R3 and R4 before they are combined. The transmitter transmits only a mono signal, created by adding the two channels. The combined signal then passes through C3, which is there to pass the audio frequencies but prevent the audio signal's average voltage of 0V from interfering with the voltage divider formed by R1 and R2. Their task is to keep the average signal voltage at one-half of the supply voltage so that both positive and negative peaks of the signal fall inside the Pico's analog signal range of 0V to 3.3V.

C2, R1, and R2 form a low-pass filter, suppressing frequencies above approximately 7kHz. The audio signal is then passed to Pin 32 of the Pico, which connects internally with one of its analog-to-digital converters, called ADC1. The trimmer potentiometer, P1, is wired to Pin 31, which connects internally with the analog-to-digital converter that you used previously, named ADC0. This adjusts the carrier frequency, depending on the position of the trimmer.

The positive supply voltage for the audio input components is taken from Pin 35, which carries the 3.3V reference voltage for the analog-to-digital converters. This pin provides current that is better regulated and less noisy

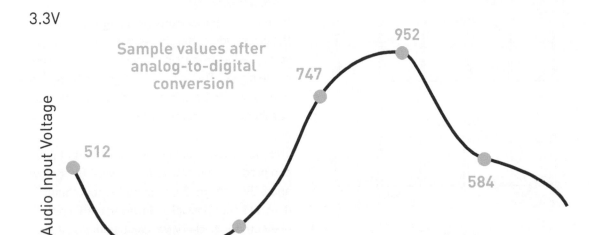

3.3V

Audio Input Voltage

Sample values after analog-to-digital conversion

512

225

307

747

952

584

6-4 Sampling an analog signal. A varying input voltage is converted to digital values at fixed intervals of time.

than the supply from Pin 36. Pin 35 cannot supply much current, but it's enough to power the voltage divider and the frequency-selection trimmer.

In the sketch in Figure 6-3, which you uploaded to the Pico, the **analogRead** function is used twice. On line 23, the statement reads the voltage of the trimmer, which controls the carrier frequency. On line 38, the statement samples the audio-input voltage to Pin 32 about 30,000 times per second. This sampling process digitizes the audio waves by converting the voltage at each instant into a number from 0 through 1,023. Samples are taken at a constant frequency, as suggested in Figure **6-4**.

Every digital device that processes sound from a microphone has to deal with sampling in some way. Once the signal has been converted to a sequence of samples, it can be processed further using mathematical operations. This is called **digital signal processing**.

When dealing with conversion of an analog signal to a digital one, there is a rule known as **Nyquist's sampling theorem**. It states that the signal must be sampled at a rate at least twice as high as the highest frequency present in the signal. In practice, it is good to have more margin, so I chose the sample rate well above twice the 7kHz threshold frequency of the low-pass filter on the input.

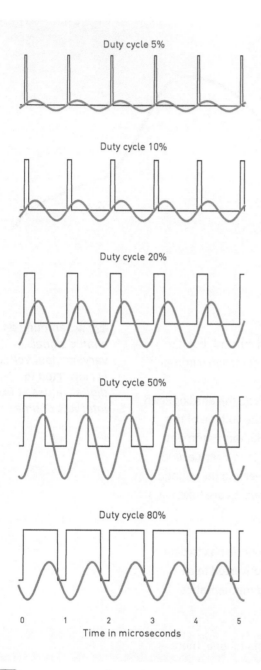

Duty cycle 5%

Duty cycle 10%

Duty cycle 20%

Duty cycle 50%

Duty cycle 80%

0   1   2   3   4   5
Time in microseconds

6-5 *A PWM signal (black) is converted by a filter so that the power in each pulse is redistributed as a smooth curve (orange). The amplitude of the output depends on the input duty cycle.*

PWM AM OUTPUT

Remember that to transmit an audio signal, the amplitude of the carrier wave—meaning its voltage—has to vary in proportion with the voltage of the signal. This is how the signal is transmitted and converted back into sound by a speaker or earphone in a radio receiver.

The Pico is a digital device, and its output pins can produce only two voltage levels: 0V for low and 3.3V for high. So, amplitude modulation doesn't seem possible. However, with a bit of electronic trickery, we can use the digital samples to change the PWM percentages of the Pico's output, and then we can use a low-pass filter to convert the square wave into a sine wave at the carrier frequency.

Figure **6-5** shows how the filtering works. The total power of each pulse, shown in black, is spread out by the filter to create a nice, smooth wave, like a sine wave, shown in orange.

The simulation shows that the filtered PWM signal is a sine wave. The frequency of it is 1MHz, like the PWM signal, and the amplitude varies according to the PWM duty cycle. So, the sketch just needs to measure the audio signal and use it to set the PWM duty cycle. The next thing to figure out is what range of duty cycles to use. You might think that a duty cycle of 100% gives the highest amplitude, but in fact, the amplitude declines after a 50% duty cycle. Therefore, the full range of the audio signal should be converted to a duty cycle between 0% and 50%.

Figure **6-6** shows how the sketch samples the audio signal with the analog-to-digital converter at a rate of about 30,000 samples per second, shown by the vertical white lines. For every sample, the PWM percentage is set according to the current audio voltage. The third panel shows the output after filtering. The result is a signal where the amplitude is varied according to the audio signal—an amplitude-modulated signal.

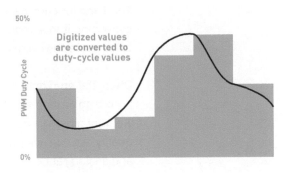

Conveniently, we figured out how to make the PWM system perform two tasks: generate the carrier frequency and enable amplitude modulation by varying the pulse duty cycle.

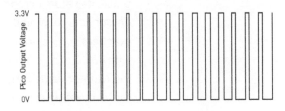

This is not a typical use of PWM, but it enables us to build a real radio transmitter using the absolute minimum number of components.

UNDERSTANDING THE SKETCH

Much of the sketch should be familiar from the previous experiment. Lines 6–11 define constants. The upper and lower limits of the tuning range are **f_min** and **f_max**. The **scale** variable acts as a volume control for the incoming audio signal. The **setup** function initializes the display as before.

Line 18 sets the output pin drive strength, controlling how much current the pin can supply. It is set to the highest value to make the signal as strong as possible. Line 19 finds the system clock frequency of the Pico and stores it in the variable **f_sys**. (The default speed is 133MHz, but it can be changed in the Arduino IDE menu, and we want the sketch to work for different frequencies.)

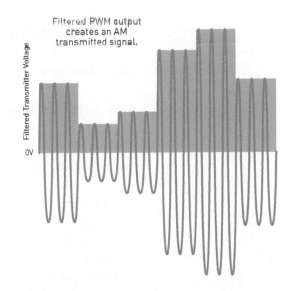

6-6 *How the Pico transforms a digitized audio input into a modulated carrier sine wave.*

The **loop** function begins by reading the trimmer voltage on line 23 and converting it to a frequency on line 24. Lines 25–28 set up the PWM system to generate the carrier wave with the frequency *f*.

Often, the **analogWrite** function takes care of setting up the PWM system according to the frequency and analog range the user specifies. This time, we want a high frequency and the best possible analog range for good audio quality; we can achieve this by requesting a good combination of frequency and analog range.

The PWM system contains a binary counter, which counts pulses from the Pico's system clock. The counter starts with the value 0, and for every clock pulse, its value increases by 1. When the counter reaches a value I'll call the **divisor**, it is reset and starts again from 0. So, the counter goes through the numbers 0, 1, 2, and so on until it reaches **divisor-1**, and then resumes with 0, 1, 2. The counter value is fed to a binary comparator, which compares the counter value with a value called **level**. If the counter value is less than the level value, the comparator output is 1; otherwise, it is 0. The idea behind all of this is that the comparator output is the PWM signal. Its frequency is

f = f_sys / divisor

and its duty cycle is

duty = 100% * level / divisor

A full description of the Pico PWM system can be found in the Pico datasheet and in the SDK manual.

Line 25 finds the integer divisor that gives a frequency as close as possible to the target *f*. Line 26 calculates the frequency actually obtained when using the integer divisor, which will be close to the target frequency. Lines 27 and 28 set up the PWM system with *f* as the frequency and the divisor as the analog range. (The divisor value is displayed on the screen when the program runs; you can verify that the system-clock frequency divided by the divisor equals the displayed frequency.)

Lines 29–34 display the frequency, the divisor value, and the peak audio amplitude in terms of percentages of the full signal range on the screen.

The actual AM signal is generated in the **for** loop from lines 37–45. The audio signal value is read and scaled to the range 0–1.

The duty-cycle value is then calculated and stored in the variable **out**. An audio value of 0 should give a duty cycle of 0%, and a value of 1 should give a duty cycle of 50%. Line 42 shows how this could be done, with the duty cycle proportional to the audio value. However, the sine amplitude is not linearly proportional to the duty cycle, and line 43 corrects for this nonlinearity, calculating the out value that results in a signal amplitude proportional to the audio signal. The function **asinf** is the inverse of the sine function, and the final **f** in the name means we are using the floating-point version of the function, and **M_PI** is a name for pi in the C language. Why the expression looks exactly like this has to do with how the amplitude of the different sine waves that make up the PWM signal (like those shown in Figure 3-10 for a square wave) depend on the pulse duty cycle. Line 44 sends the calculated duty cycle to the output pin.

Now that you know how the sketch works, it is time to look at improving the circuit for a longer transmission range.

LONGER RANGE AND FILTERING

You may object that while I show pictures of filtered PWM signals, there is no filter on the transmitter output pin in the schematic. This is true! However, the receiver performs some filtering. Tuning the receiver to a certain frequency means that the receiver will only pick up signals in a narrow interval around that frequency.

However, this doesn't mean that transmitting an unfiltered square wave is a good idea. As I explained in Experiment 3, a square wave signal contains overtones or harmonics, so you are in fact transmitting on multiple frequencies simultaneously. This may lead to interference with other radio transmitters. The reason I felt comfortable with that experiment is that the transmission range with the small wire loop is very short.

Now, our objective is to increase the range, which means we have to do the job right and add filtering to AMT2B. We can achieve this by designing a better antenna—which also, conveniently, acts as a filter.

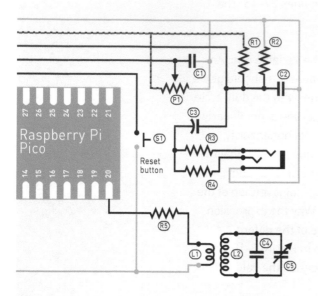

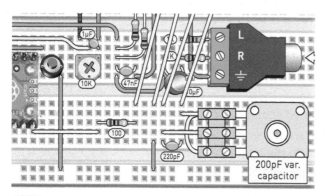

6-7 *Adding coils and capacitors to create a better antenna. The left side of the circuit is not shown, because it is unchanged.*

6-8 *The breadboarded version of the circuit in Figure 6-7.*

The idea is to use a resonance circuit and tune it to the transmission frequency. The coil of the resonance circuit acts as the antenna, and because the circuit is tuned to resonate only with the intended transmission frequency, the harmonics will be filtered out. The result is a cleaner signal and a longer transmission range.

CONSTRUCTION

The schematic for the new antenna circuit is shown in Figure **6-7**, and the breadboarded version is shown in Figure **6-8**. The upper part of the circuit remains the same, but the antenna loop has gone and is replaced by a coil named L1. Next to it is L2, which is connected to a tuning capacitor. Before you start the construction, set up the transmitter and receiver, and tune them to the same frequency, near 900kHz.

Then, disconnect the transmitter from the computer for the following steps, and power off the receiver to save batteries.

Start by winding the coils. They should both have the same diameter of approximately 3.5" so you can wind them around the same base object. You'll need a nonconductive and nonmagnetic cylindrical object for this, such as a plastic jar or bottle. A piece of PVC pipe for plumbing work may also be suitable. The first coil has 2 turns and the second has 27. Start with the coil consisting of 2 turns, with a 6" tail at each end. Twist the tails together and strip the insulation from the ends for breadboard connection. Repeat the procedure for the coil of 27 turns, winding it in the same direction, adjacent to the previous one, with no space between them. Secure the two coils with tape if needed. The actual coils are shown in

6-9 *The coils to improve transmission power of AMT2B.*

Figure **6-9**. They are not opposite to each other, as shown in the schematic; that's just the way a schematic indicates that two coils are interacting with each other.

Connect the 2-turn coil to the transmitter output, as shown in Figures 6-7 and 6-8, instead of the wire loop that you used previously. The tails of the 27-turn coil also connect to the breadboard, across C4, a 220pF capacitor. This is wired in parallel with C5, the same type of variable capacitor that you used previously for AMR2. (If you want to receive the transmission using AMR2, that will still require its own variable capacitor.)

Reconnect the transmitter to your audio source, and power on the receiver. Keep the transmitter and the receiver tuned to the same frequency as before. Adjust the tuning capacitor of the transmitter, and you should find a setting where the receiver picks up a strong signal. Find the setting that gives the strongest signal. You have now tuned the antenna-resonance circuit to the frequency of the transmitter.

If you want to tune the transmitter to the high end of the AM band, you may have to remove C4.

Try to move the receiver away from the transmitter until the signal becomes too weak to receive. Has the transmission range increased?

Adding the resonance circuit to the output has two effects. When the frequency of the resonance circuit matches the frequency of the transmitter, it resonates, and the current going back and forth through L2 becomes large, meaning that the magnetic field from the two coils is stronger than the field from L1 on its own. Also, the resonance circuit does not react to the harmonics present in the signal, so they will not be amplified.

A drawback with this resonance circuit is that to change the transmission frequency, there are now two controls to tune: the transmitter frequency and the antenna-resonance frequency. But it's common for antennas to work best when they are resonant with the frequency they transmit or receive, and this may involve some tuning.

Winding 27 turns seems like a lot. The reason we need this many turns is that the coil and the capacitor should be able to resonate with a frequency in the broadcast band, and this needs a large inductance. With a 100µH coil and 420pF capacitance (which we get from the 220pF ceramic capacitor plus the 200pF variable capacitor set to its maximal value), the frequency is 776kHz. Higher frequencies can be tuned up to about 1MHz by turning the variable capacitor and then disconnecting the 220pF capacitor for even higher frequencies.

We could have used a smaller inductance and larger capacitance, but then the tuning range achievable with the variable capacitor would be impractically small.

With a larger coil diameter, you would need fewer turns and should also get a longer transmission range. However, the number of turns you need decreases more slowly than the diameter increases, so you end up needing more wire for a larger coil. As a rule of thumb, if you double the coil diameter, you can divide the number of turns by the square root of 2, about 1.4, to keep the inductance constant. To calculate the inductance of a coil you design yourself, you can use the Wheeler approximation (described in Experiment 1) or search online for "coil-inductance calculator."

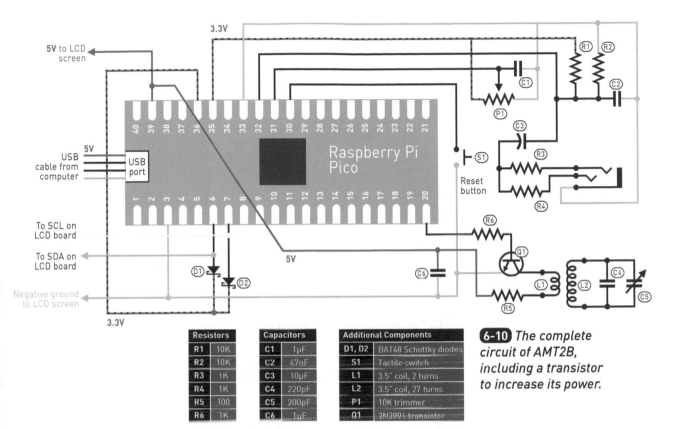

Resistors		Capacitors		Additional Components	
R1	10K	C1	1μF	D1, D2	BAT48 Schottky diodes
R2	10K	C2	47nF	S1	Tactile switch
R3	1K	C3	10μF	L1	3.5" coil, 2 turns
R4	1K	C4	220pF	L2	3.5" coil, 27 turns
R5	100	C5	200pF	P1	10K trimmer
R6	1K	C6	1μF	Q1	2N2904 transistor

6-10 *The complete circuit of AMT2B, including a transistor to increase its power.*

MORE POWER?

What if you want to extend the range of the transmitter a bit more, while of course staying within the legal limits? With the square wave harmonics eliminated, it is now possible to amplify the signal with a transistor. The circuit in Figures **6-10** and **6-11** on the following page shows how it can be done. Notice that R5 has changed its position, although it still has the same function, limiting current through L1. R6 has been added to limit current through the base of transistor Q1. The complete list of parts in this final version of the circuit is shown in Figure 6-10.

Disconnect the transmitter from the audio source and your computer before making changes. Don't change the frequency setting or the tuning capacitor, to keep the radio tuned for the next test. When you have assembled and checked the circuit, reconnect the USB cable and the signal source. Transmit again, and see if the range has increased.

The transistor amplifier is very simple, because it is only dealing with the digital output from the Pico, which is either on or off. This also makes the amplifier efficient, because the transistor either conducts fully or not at all,

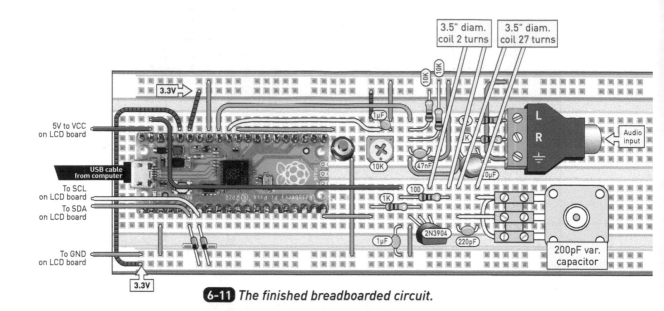

6-11 *The finished breadboarded circuit.*

and in both cases, the amount of electric power lost to heating the transistor is small.

In an analog amplifier where a transistor receives a variable input and delivers a variable output, the power loss in the transistor can be considerable, equal to or exceeding the power delivered to the amplifier output. For this reason, many modern, high-power broadcast AM transmitters use some form of pulse-width modulation.

When the Pico output is high, the transistor conducts and grounds one end of coil L1 so that current flows through it. When the Pico output is low, the transistor doesn't conduct, and no current flows in L1. L2, C4, and C5 form a resonance circuit as before. R5 limits the current, both to protect the transistor and to ensure that the signal stays in the allowed power.

I mentioned the regulations affecting transmitters in the United States in Experiment 3. You are allowed a maximum of 100mW power and an antenna of no more than 10 feet. In this case, the power limit is the total power supplied to the transistor. I calculated this to be around 0.1W.

A further advantage with adding the transistor amplifier is that the audio quality is improved. The unamplified transmitter is somewhat noisy. I think the noise enters the signal in the analog-to-digital converter and is related to the high-frequency PWM signal the Pico creates. With the transistor in place, the Pico has to handle and output less current, which leads to less interference.

This is a common problem in circuits mixing digital and analog sections. Noise from digital switching enters the analog signals. I tried different modifications to reduce the noise. Adding capacitors between ground and the different positive voltages, while generally a good idea, gave little improvement. What did help was to use a separate ground pin on the Pico for the audio and the trimmer section of the circuit. Pico Pin 33, AGND, is designated as analog ground. It is otherwise similar to the other ground pins, but the circuit board is designed so that the AGND and analog input traces go together for minimal interference from other circuitry.

If you want to use the transmitter on a more permanent basis to transmit music to an AM radio in your house, you can find a frequency free from other stations and tune your transmitter there permanently. Modify line 24 in the sketch to set f to the frequency you want, in hertz. Then you avoid the small shift in audio quality that happens if the transmitter shifts slightly in frequency as a result of the trimmer reading changing slightly. Or you can adjust **f_min** and **f_max** closer to your chosen frequency. Then you still have the possibility to tune the frequency a little, but the tuning will jump less.

STATUS UPDATE

In this experiment, you saw how the Pico can run fast enough to disassemble an audio signal into a series of digital values, performing the process known as analog-digital conversion. Then you saw how the digitized signal could be adapted using digital signal processing, and how filtering can convert square waves into sine waves. Finally, the power of the transmitter was increased with transistor amplification.

The next step is to use additional capabilities of the Pico to do even more, as it counts the pulses in a carrier wave.

7

COUNTING PULSES WITH A PICO

When building the AM transmitter in Experiment 3 and the test signal generator in Experiment 1, the only way to determine the frequency of a carrier wave was by doing some calculations using the values of coils, capacitors, and resistors. Then you could check the math, approximately, by trying to find the signal with an AM radio tuning dial.

Anytime you transmit a radio signal, though, you really need a more precise way to measure its frequency.

Your digital multimeter may allow this as an option, but is probably limited to a few kilohertz. In this experiment, the Pico can measure frequencies up to about 60MHz, displaying the result on the LCD screen that you have been using in the last three experiments.

As you learned in Experiment 1, the frequency of a signal is measured in hertz, the number of pulses per second. Conveniently, it is very easy for the Pico to count pulses during a one-second interval.

You Will Need:

- The Pico on a breadboard, hooked up to an LCD screen (1), as has been featured in Experiments 4, 5, and 6.
- Optional: Transmitters AMT0 from Experiment 1 or AMT1 from Experiment 3.
- Ceramic capacitors: 100pF (1), 10nF (1), 1µF (1).
- Resistors: 1K (1), 2.2K (2).
- 7555 integrated circuit chip (1).
- BAT48 Schottky diodes (4). Two of these diodes have been used in previous experiments. Two more will be needed in this experiment.

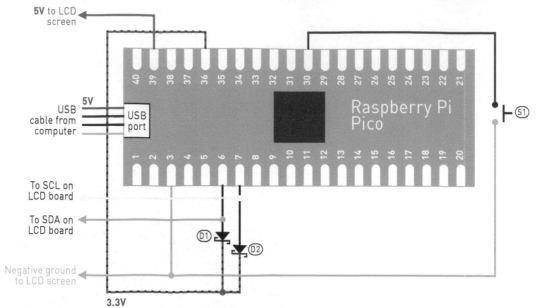

5V to LCD screen

USB cable from computer

5V

USB port

To SCL on LCD board

To SDA on LCD board

Negative ground to LCD screen

3.3V

Raspberry Pi Pico

7-1 *The first step toward building a frequency counter.*

```
 1 #include <Wire.h>
 2 #include <hd44780.h>
 3 #include <hd44780ioClass/hd44780_I2Cexp.h>
 4 #include "hardware/pwm.h" // pico-specific PWM functions
 5 hd44780_I2Cexp lcd;
 6 const int LCD_COLS = 16, LCD_ROWS = 2;
 7 const int input_pin = 15;
 8 const int output_pin = 16;
 9 const int divisor = F_CPU / 10000;   // 10 kHz test signal
10 int slice_num = pwm_gpio_to_slice_num(input_pin);
11 int output_slice_num = pwm_gpio_to_slice_num(output_pin);
12
13 void setup()
14 {
15    int status = lcd.begin(LCD_COLS, LCD_ROWS);
16    if(status)
17      hd44780::fatalError(status);
18    gpio_set_function(output_pin, GPIO_FUNC_PWM); // test signal
19    pwm_set_wrap(output_slice_num, divisor-1);
20    pwm_set_gpio_level(output_pin, divisor/2);
21    pwm_set_enabled(output_slice_num, true);
22    gpio_set_function(input_pin, GPIO_FUNC_PWM);  // input
23    pwm_set_clkdiv_mode(slice_num, PWM_DIV_B_RISING);
24 }
25 void loop()
26 {
27    pwm_set_counter(slice_num, 0);       // reset counter
28    pwm_set_enabled(slice_num, true);    // start counting
29    delay(1000);
30    pwm_set_enabled(slice_num, false);   // stop counting
31    int count = pwm_get_counter(slice_num);
32    lcd.clear();
33    lcd.print(count);
34    lcd.print(" Hz");
35 }
```

7-2 *A first version of the frequency-counter sketch.*

MEASURING FREQUENCY

Here's what your Pico has to do:

- Start to count arriving pulses on an IO pin.
- Measure an interval of one second.
- Display the count so far.

To count pulses, you could write a program to repeatedly check the state of the pin, and every time it sees a transition from low to high, it would increment a counter variable. For low frequencies, this would work. But what if the frequency is high and the program also has other work to do, such as keeping track of the time? The Pico could miss a pulse, in which case the measurement would be wrong, and you would have no way of knowing. A better way is to use a dedicated hardware counter, which is built into most microcontroller chips, including the Pico. I will show you how to use the PWM system for this purpose.

CONSTRUCTION

Construct the circuit shown in Figure **7-1**, using the basic Pico circuit that we established previously. All you need to do is remove most of the components.

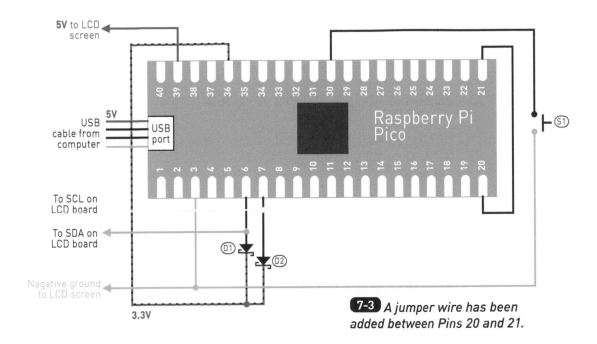

5V to LCD screen

USB cable from computer

5V

USB port

Raspberry Pi Pico

To SCL on LCD board

To SDA on LCD board

Negative ground to LCD screen

3.3V

7-3 *A jumper wire has been added between Pins 20 and 21.*

Now, upload the "7-1-frequency_counter" sketch in Figure **7-2**, which you can copy from my GitHub page at github.com/ fjansson/MakeRadio. You should see the screen display

0 Hz

How can you test it to make sure it is working and accurate? You can use the Pico itself to output a known frequency on another pin and measure that. The sketch you just uploaded can do this: It configures Pin 21 as a PWM output and outputs a 10kHz signal there. You just need to add a jumper wire from that pin to the counter input, which is on Pin 20, as shown in Figures **7-3** and **7-4**. Now, the LCD screen should show you

10000 Hz

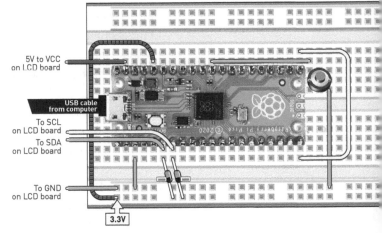

5V to VCC on LCD board

USB cable from computer

To SCL on LCD board

To SDA on LCD board

To GND on LCD board

3.3V

7-4 *The breadboarded version of Figure 7-3.*

How would you like to measure a higher frequency, such as 1MHz? You can try this by editing line 10 in your sketch to

```
const int divisor = F_CPU / 1000000;
```

Upload the sketch again. Now, the screen reads

16960 Hz

But that's not right! Perhaps the sketch doesn't handle high frequencies yet. To understand why, and how to improve the sketch, we need to look at how the sketch works, and learn about programming using *interrupts*.

HOW THE SKETCH WORKS

In this sketch, the Pico's PWM hardware is used to count pulses arriving on the input and to generate a test signal. In Experiments 5 and 6, you used the Arduino function **analogWrite** to control the PWM system. In this experiment, you have to use functions specific to the Pico because the Arduino PWM functions deal with output but not with counting pulses. The **include** command on line 4 allows the sketch to use the Pico's own functions for controlling the PWM system.

Lines 6–9 define constants for the sketch. **input_pin** is the GP pin number of the counter input, **output_pin** is the test signal output, and **divisor** sets the frequency of the test signal. The output frequency is the system clock's frequency divided by the divisor.

The Pico contains eight identical PWM blocks, also called *slices*, with two outputs each. Lines 10 and 11 find which PWM blocks are associated with the input and output pins.

One of the PWM blocks is shown in Figure **7-5**. Each one contains an 8-bit counter called a *prescaler*, a 16-bit *counter*, and two *comparators*. The prescaler can be used to divide the clock frequency by a number between 1 and 256. In this experiment, it is not used, so clock pulses will pass through it (which is the same as dividing the incoming frequency by 1). For the task of generating PWM signals, the 16-bit counter counts pulses from the system clock (as long as the switch at the left-hand side of Figure 7-5 is in the upper position). When the counter reaches the value wrap, it starts again from 0 at

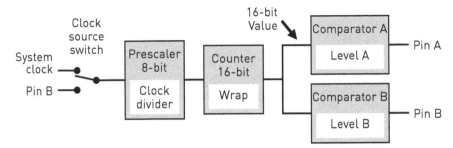

7-5 *The internal structure of one of the eight PWM blocks in the Pico. Output A and Output B can be routed to Pico pins. The clock input at the left can come either from the system clock (for PWM output) or from Pin B (used as an input for counting pulses).*

the next pulse. Each of the comparators compares the binary value from the counter with the level value for each output. If the counter value is less than the level value, the output is high. If not, the output is low. In other words, the duty cycle of the output signal is set by the level value. The frequency of the output signal is the system-clock frequency divided by wrap plus 1.

Lines 18–21 set up the test-signal output. Line 18 assigns the output pin to its PWM block. Line 19 sets the output frequency using the divisor constant, and line 20 sets the duty cycle to 50%. Finally, line 21 turns on the PWM output.

To use the PWM block to count pulses on an input pin, the switch at the left of Figure 7-5 is flipped to the lower position. Line 22 sets the input pin connected to the PWM system, and line 23 instructs the PWM block associated with the input pin to count rising edges on the input pin. This means that every time the input pin changes state from low to high, the counter increments.

In the **Loop** function, line 27 sets the counter value to 0 to prepare for a measurement. Line 28 enables the counter so that it increments for every low-to-high transition on the input. Line 29 waits for 1 second, and line 30 disables the counter again. Now, the counter contains the number of pulses that arrived during the one-second delay. This number is retrieved on line 31 and displayed on the LCD with lines 32–34.

```
1  #include <Wire.h>
2  #include <hd44780.h>
3  #include <hd44780ioClass/hd44780_I2Cexp.h>
4  #include "hardware/pwm.h" // pico-specific PWM functions
5  hd44780_I2Cexp lcd;
6  const int LCD_COLS = 16, LCD_ROWS = 2;
7  const int input_pin = 15;
8  const int output_pin = 16;
9  const int divisor = F_CPU / 1000000;    // 1 MHz test signal
10 int slice_num = pwm_gpio_to_slice_num(input_pin);
11 int output_slice_num = pwm_gpio_to_slice_num(output_pin);
12 volatile int counter_wraps;
13
14 void pwm_interrupt_handler()
15 {
16    pwm_clear_irq(slice_num);
17    counter_wraps++;
18 }
19 void setup()
20 {
21    int status = lcd.begin(LCD_COLS, LCD_ROWS);
22    if(status)
23       hd44780::fatalError(status);
24    gpio_set_function(output_pin, GPIO_FUNC_PWM); // test signal
25    pwm_set_wrap(output_slice_num, divisor-1);
26    pwm_set_gpio_level(output_pin, divisor/2);
27    pwm_set_enabled(output_slice_num, true);
28    gpio_set_function(input_pin, GPIO_FUNC_PWM); // input
29    pwm_set_clkdiv_mode(slice_num, PWM_DIV_B_RISING);
30    pwm_clear_irq(slice_num);
31    pwm_set_irq_enabled(slice_num, true);
32    irq_set_exclusive_handler(PWM_IRQ_WRAP, pwm_interrupt_handler);
33    irq_set_enabled(PWM_IRQ_WRAP, true);
34 }
35 void loop()
36 {
37    counter_wraps = 0;
38    pwm_set_counter(slice_num, 0);       // reset counter
39    pwm_set_enabled(slice_num, true);    // start counting
40    delay(1000);
41    pwm_set_enabled(slice_num, false);   // stop counting
42    int count = pwm_get_counter(slice_num);
43    lcd.clear();
44    lcd.print(counter_wraps * 65536 + count);
45    lcd.print(" Hz");
46 }
```

7-6 *Final frequency counter sketch, using interrupts to detect when the counter wraps around.*

The sketch is quite simple, since the hardware counter is doing the work of recognizing incoming pulses and counting them. The only difficult thing in the sketch is setting up the PWM system. The RP2040 datasheet and the Pico SDK manual explain how the PWM system is constructed and which C functions are available for using it, and show how the IO pins are associated with the eight PWM blocks.

HIGHER FREQUENCIES

Now, we can come back to the question of why the counter failed with the 1MHz input signal. If the counter runs for one second and pulses arrive at a rate of 1MHz, it should have received one million pulses.

Look back at Figure 7-5. There, you see that the main counter is 16-bit, meaning it can hold numbers from 0 to $2^{16} - 1 = 65,535$. After this amount of pulses, the counter starts again from 0, and this is what happens with the 1MHz signal.

So, what do we do? We could let the counter run for a shorter time—perhaps 10ms. But then the accuracy would suffer for low frequencies, or we'd have to make the time interval adjustable. Instead, we'll extend the counting range using interrupts.

INTERRUPTS ON THE PICO

Interrupts are a mechanism for computers to handle events: "An important thing happened! Interrupt the program you are running and run this function instead. Once you are done, go back to the normal program." Interrupts are useful on microcontrollers, and typically, you can set up interrupts for many things. For example:

- A timer reaches the end of its interval.
- An input pin changes its state—for example, because a button was pressed.
- The PWM counter wraps around to 0.
- A character is received on a serial port (or USB).

That third example—the PWM counter wrapping around to 0—is the situation we need to address with an interrupt. Each time the interrupt occurs, a function is called. In this function, a second counter is incremented that keeps track of how many times the first counter wrapped around. The total number of pulses can be calculated from the values of the two counters.

The final "7-2-frequency_counter" sketch, which you can copy from my GitHub page at github.com/fjansson/MakeRadio, is shown in Figure **7-6**. Load it into the IDE to replace the previous version.

On line 12, the new counter variable is defined. The keyword **volatile** indicates to the compiler that the variable's value can change unexpectedly (because the value is modified when interrupts occur).

Line 14 defines a new function, **pwm_interrupt_handler**. This function is called whenever the PWM interrupt occurs. The **pwm_interrupt_handler** function itself does very little; on line 17, it increments the variable **counter_wraps**, keeping track of how many times the PWM counter has wrapped around. On line 16, it calls **pwm_clear_irq(slice_num)** to clear the interrupt-request flag associated with this interrupt. Clearing the flag marks the interrupt as handled so that the interrupt handler function doesn't immediately get called again. If you omit this line, you will notice because the Pico locks up. (Hold down the **BOOTSEL** button on the Pico board, while pressing and releasing the **Reset** button on the breadboard to make it communicative again, as explained in Experiment 4.)

In the **setup** function, lines 30–33 are new: They set up the PWM interrupt. Line 30 clears any pending interrupts, line 31 enables interrupts for the specific PWM block used for counting, line 32 associates the **pwm_interrupt_handler** function with the PWM interrupt, and line 33 enables interrupts for the PWM blocks in general.

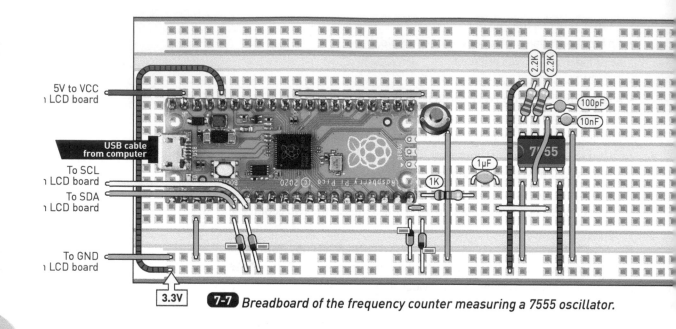

5V to VCC
ı LCD board

USB cable
from computer

To SCL
ı LCD board

To SDA
ı LCD board

To GND
ı LCD board

3.3V

7-7 *Breadboard of the frequency counter measuring a 7555 oscillator.*

In the loop function, line 37 resets **wrap_counter**, and the code on line 44 for displaying the result has been modified to calculate the number of pulses using the two counters.

With this sketch, we can measure frequencies up to half the Pico's system-clock speed. (Because of how the PWM system is constructed, it needs to see the signal being high for at least one system-clock cycle and low for at least one.) When we measure the 1MHz test signal, the display now reads

1000001 Hz

Perhaps the time interval is slightly too long because of the delay function or because of the few instructions it takes to enable and disable the PWM counter. Anyway, the accuracy seems quite good!

Speaking of accuracy, remember that the test signal was generated with the same Pico that did the measurement. How accurate is it, really? Both the test signal and the time interval for counting pulses are determined by a certain number of clock cycles of the system clock. This clock is controlled

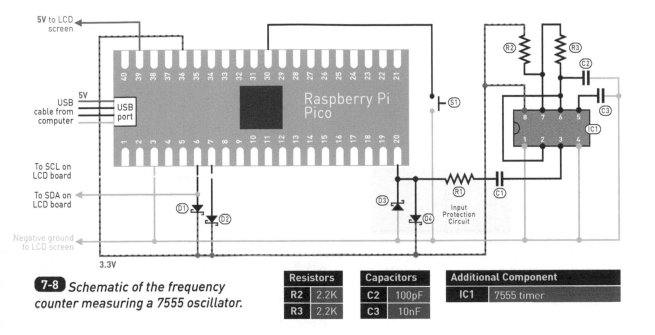

7-8 Schematic of the frequency
counter measuring a 7555 oscillator.

Resistors	
R2	2.2K
R3	2.2K

Capacitors	
C2	100pF
C3	10nF

Additional Component	
IC1	7555 timer

by a 12MHz quartz-crystal oscillator, which you can see in Figure 4-11 as
a small rectangular metal box between Pins 14 and 27. This oscillator is
specified to have an accuracy of 30 parts per million. The frequencies we
generate with the PWM system have the same relative accuracy, so a 1MHz
signal is supposed to be accurate to within 30Hz. You can try increasing the
test frequency further.

MAKING MEASUREMENTS

Now that the counter is working and has been tested at high and low
frequencies, you can test it on a real circuit. Remove the jumper wire that
you added between Pins 20 and 21, and assemble the circuit shown in
Figures **7-7** and **7-8**.

You may recognize the components around the 7555 timer as being the
same as the ones you used in Experiment 2 to test receivers. The 7555 is
able to run on a 3.3V power supply, which fits the Pico circuit nicely.

Apply power to the circuit by connecting the USB cable to your computer.
You should see a frequency between 800kHz and 900kHz displayed on

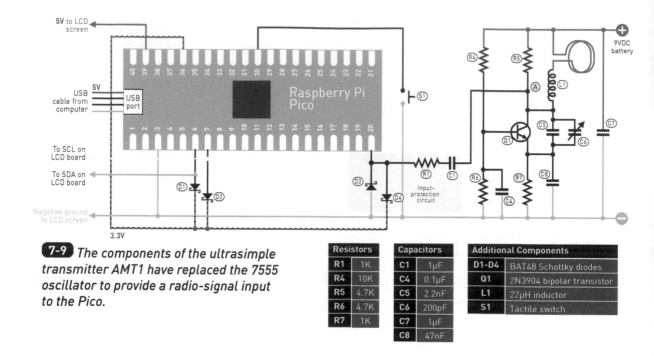

5V to LCD screen

USB cable from computer

5V

USB port

To SCL on LCD board

To SDA on LCD board

Negative ground to LCD screen

3.3V

Raspberry Pi Pico

Input-protection circuit

9VDC battery

7-9 *The components of the ultrasimple transmitter AMT1 have replaced the 7555 oscillator to provide a radio-signal input to the Pico.*

Resistors	
R1	1K
R4	10K
R5	4.7K
R6	4.7K
R7	1K

Capacitors	
C1	1µF
C4	0.1µF
C5	2.2nF
C6	200pF
C7	1µF
C8	47nF

Additional Components	
D1–D4	BAT48 Schottky diodes
Q1	2N3904 bipolar transistor
L1	22µH inductor
S1	Tactile switch

the screen. The frequency will fluctuate between measurements, and if you increase the temperature of the 7555 circuit by blowing on it, you can see the frequency change by up to 1kHz. The values of the resistors and the capacitor that determine the frequency change slightly when the temperature changes, and the properties of the timer chip itself change as well. This poor frequency stability is a reason 555 and 7555 timers are not often used in radio applications—the stability of quartz crystals, such as the one keeping time in the Pico, is much better.

INPUT CIRCUIT

In Figure 7-8, the components in the blue rectangle will protect the Pico's input from high voltages. They allow you to measure signals with an amplitude larger than the Pico's voltage range of 0V to 3.3V. This is necessary in our next modification of the circuit to measure the frequency of the AMT1 transmitter from Experiment 3.

In Figures **7-9** and **7-10**, the components of AMT1 have been inserted where the 7555 and its associated components had been. Because the transmitter requires a 9V power supply, a battery has been added to the positive bus along the upper edge of the breadboard. The negative side of the battery shares the negative bus at the bottom of the breadboard.

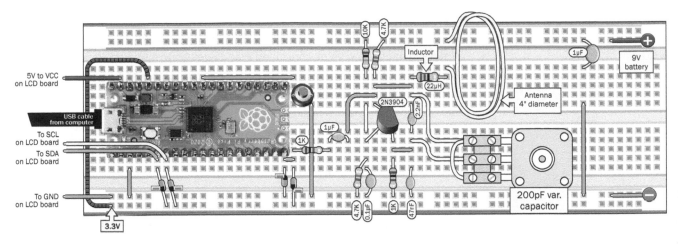

7-10 *The breadboarded version of Figure 7-9.*

The voltage at point A in the transmitter fluctuates around an average value of about 8V, and that voltage can damage the Pico input. Capacitor C1 in the counter lets voltage fluctuations through and blocks the constant part of the signal.

The next problem is that the signal amplitude in Figure 3-3 is 12V peak to peak, too much for the Pico. We have placed Schottky diodes D3 and D4 from the input pin to ground and to the 3.3V supply (like the I2C lines in Experiment 4), but now, we also need to protect against voltages below the ground level. Note the polarity of the diodes. They are turned so that the diode to the +3.3V bus conducts if the pin voltage is higher than 3.3V and the diode to ground conducts when the pin voltage is below 0V. Resistor R1 is in the circuit to limit the current through these diodes. The capacitor, the 1K resistor, and the two Schottky diodes form the input circuit of the counter.

MEASURING AMT1

Don't connect any audio signal to the input of AMT1. You should see the frequency of the transmitter's carrier wave on the LCD screen, probably in the range 750kHz to 800kHz. If you turn the tuning capacitor, you should see the frequency change. Even when you don't touch the circuit, you can see the count fluctuating. If you blow on the transistor to warm it, you can also see the frequency slowly shift. You are seeing one of the weaknesses of AMT1—as you saw for the 7555 timer, its frequency stability is not very good. When you used the crystal-controlled Pico to generate signals—in both this experiment for the 1MHz test signal and in Experiments 5 and 6 with the trimmer-controlled signal generator—the stability is much better. (I have to mention, though, that LC oscillators such as the one in AMT1 can be made quite stable when constructed carefully and shielded from outside influence. Before quartz crystals became available, LC oscillators were commonly used in radio receivers and transmitters.)

You now have a frequency counter that is accurate and can deal with frequencies up to 15MHz. What the counter still cannot do is deal with weak signals. You could add an amplifier for that. But for the projects in this book, whenever we have an oscillator where we want to measure the frequency, the signal amplitude will be sufficient.

STATUS UPDATE

This experiment provided some tips and tricks for measuring the frequency of a carrier wave, and included the code that a Pico needs to perform that task. This takes us as far as we need to go into the theory of AM radio. Amplitude modulation is not the only way to transmit audio signals to a receiver. FM radio was patented back in 1933, and became the dominant system for broadcasting music. In the next experiment, I'll show you how FM stereo broadcasts work, and you'll learn how to build a receiver using the Pico.

experiment

8

FM RECEIVER

All of the transmissions you have created and received so far have used amplitude modulation, abbreviated as *AM*. But thanks to the work of radio pioneer Edwin Armstrong, we can all enjoy better signal quality from frequency modulation, known as FM.

Today, most radio stations that broadcast music use FM. In this experiment, you'll be able to pick up those transmissions with your Raspberry Pi Pico by adding a plug-in FM module, a little circuit board measuring about 1" × 1".

You'll need a new sketch in order to do this, and I'll take you step by step through the code to help you understand how it works.

You will also learn how to power the Pico with a battery so that you can disconnect it from your computer and carry it around as a portable radio. Finally, you'll get some recommendations for more resources to take you deeper into the capabilities of the FM module.

You Will Need:

- FM receiver module (1). SparkFun WRL-12938, HiLetgo Si4703, or similar breakout board containing an Si4703 FM chip.
- Pin header to fit WRL-12938, male, through-hole, 8 pins, single row, 0.1" spacing, TE Connectivity 5-146282-8, Adafruit 5584 short plug headers or similar (1).
- Soldering iron and solder to mount the pin headers on the FM receiver module board.
- Raspberry Pi Pico and LCD screen, as used in previous experiments (1).
- Electrolytic capacitor, 470µF (1).
- Ceramic capacitor, 1µF (1).
- BAT48 Schottky diodes (2).
- Any stereo headphones or earphones with 1/8" plug (1).
- Tactile button switches (3).

The next three items are optional if you decided to add speakers:

- Speakers, 8 ohms, approximately 4" diameter (2).
- 1/8" audio jack with screw terminals, as used in previous experiments (1).
- Audio cable with 1/8" plugs on both ends.

FREQUENCY AND RANGE

While FM enables better quality, it does not allow the same long range that AM can achieve. This is related to its use of the VHF (very high frequency) waveband.

Most countries use the VHF range of 87.5MHz to 108MHz for FM transmissions. (Japan uses 76MHz to 95MHz.) This is high above the AM range of 540kHz to 1,600kHz. It is also higher than the shortwave radio band of 3MHz to 30MHz.

It just so happens that signals in the AM band can bounce between the surface of the Earth and the Heaviside layer, in the upper atmosphere, enabling a very long range. But radio waves in the VHF band are generally not reflected in the atmosphere, and consequently, they cannot propagate beyond the horizon. This does have one advantage: Broadcast antennas for different radio stations can be located as close as 100 miles from each other, sharing the same frequency without interfering with each other, because of their limited range. Consequently, FM radio is ideal where a large number of listeners are clustered around an antenna, such as in an urban area.

SI4703 FM RECEIVER MODULE

High frequencies can be transmitted and received by short antennas. One undesirable consequence of this is that lengths of wire in a circuit, or conductors inside a breadboard, can easily interfere with each other electrically. Generally speaking, for this reason, breadboards are not suitable for building FM circuits.

Another problem with FM signals is that they are more difficult to decode, especially if you want to transmit or receive in stereo. Therefore, instead of a receiver constructed from discrete components, I suggest you use a board containing a Si4703 chip that is already optimized as an FM receiver. Because this is a surface-mount chip, it is sold for makers preinstalled on a little breakout board that's about 1" × 1", fitted with a ⅛" jack socket, as shown in Figure **8-1**. This board is designed by SparkFun (model WRL-12938), although you can find imitation boards (such as the HiLetgo Si4703) that have the same function but cost less.

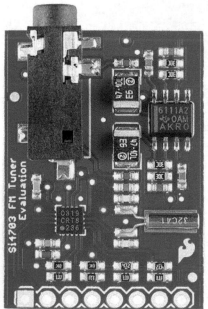

8-1 The WRL-12938 breakout board, manufactured by SparkFun and sold by numerous retailers, incorporates an Si4703 FM radio receiver chip.

The SparkFun board requires you to solder in a "header" consisting of eight pins (which is included in the above parts list). Some of the imitation boards have pins preinstalled, and you can look for them by using the search term

Si4703 board with pins

Figure 8-2 shows the pins inserted into the board and held in place with some blue masking tape, which doesn't leave a residue. I chose to solder the pins with the board upside down so that the abbreviations identifying the pins will be visible after the board has been plugged into the breadboard.

Figure 8-3 shows the pins after being soldered. You may find the soldering process is easier if you smear a little flux onto the board before you begin. You only need to apply the tip of the soldering iron for three to four seconds to each pin before you add just enough solder to make a joint.

If you use one of the imitation boards, check the location of the ground and power supply pins carefully using the labels on the back of the board. They may be different from the pin functions on the SparkFun board in my photographs.

Note that when you plug a headphone cable into the jack mounted on the board, it will also work as an antenna.

8-2 *Pins held in place with blue masking tape, ready for soldering.*

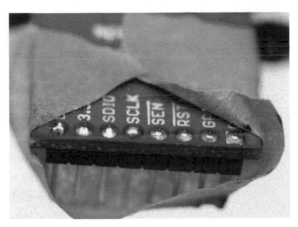

8-3 *Pins soldered into the breakout board.*

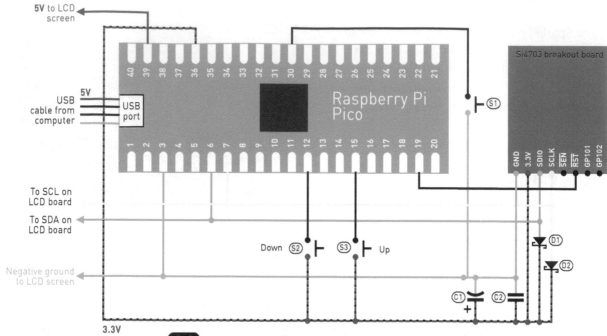

8-4 *Schematic of the FM receiver. The Si4703 module is shown here with the component side down. Note that the 3.3V and GND pins may be placed differently on your module.*

CONSTRUCTION

A schematic showing how the FM breakout board connects with the Pico is shown in Figure **8-4**. The Pico is necessary because it controls the FM breakout board, using a connection known as the *I2C bus*. This capability is built into the Pico.

Figure **8-5** shows the breadboard layout. The connections between the Pico and the LCD display are the same as in our previous experiments.

We have moved the protective Schottky diodes to make room for the two new I2C wires connecting the Pico to the radio module. Pushbuttons S2 and S3 have also been moved beyond the Pico board because they cannot fit directly between Pins 12 and 15 and the 3.3V bus on the breadboard.

The two capacitors between the positive and negative power bus on the breadboard are to reduce noise from the power supply because, as it turns out, the receiver circuit is sensitive to noise.

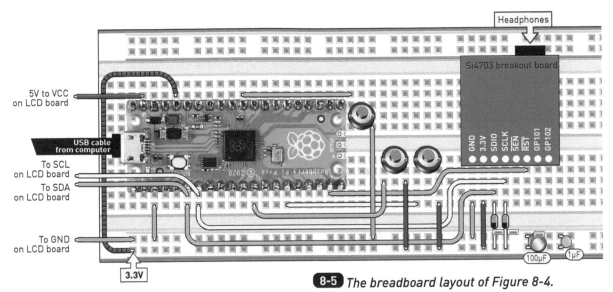

8-5 The breadboard layout of Figure 8-4.

Labels on the breadboard diagram:

Headphones

Si4703 breakout board

5V to VCC on LCD board

USB cable from computer

To SCL on LCD board

To SDA on LCD board

To GND on LCD board

3.3V

GND 3.3V SDIO SCLK SEN RST GP101 GP102

100µF   1µF

Next, you need to install a library for the Si4703 chip in the Arduino IDE. The procedure is similar to what you did in Experiment 4 with the LCD library, and you only need to do this once:

- Open the Arduino IDE on a computer with internet access.
- Open the Library Manager (Tools > Manage Libraries).
- Search **PU2CLR SI470X** and install the **SI470X** library by Ricardo Lima Caratti. (Incidentally, PU2CLR is an amateur-radio call sign, identifying a Brazilian amateur-radio station.)

Connect the Pico to your computer with a USB cable and upload the sketch in Figures **8-6a** and **8-6b** (on the following page). Plug headphones into the jack on the breakout board, which is your FM receiver. Briefly press one of the two tuning buttons (S1 or S2) to search for a station. The search should automatically stop when you find a signal, and you should see the frequency displayed on the LCD screen, with a number

```
1  #include <Wire.h>
2  #include <hd44780.h>
3  #include <hd44780ioClass/hd44780_I2Cexp.h>
4  #include <SI470X.h>
5  hd44780_I2Cexp lcd;
6  SI470X rx;
7  const int LCD_COLS = 16;
8  const int LCD_ROWS =  2;
9  const int seek_down_pin =  9; // seek down button, pin 12
10 const int seek_up_pin   = 11; // seek up button, pin 15
11 const int reset_pin     = 14; // Si4703 reset, pin 19
12 const int SDA_pin       =  4; // I2C data, pin 6
13
14 void setup()
15 {
16   rx.setup(reset_pin, SDA_pin);
17   int status = lcd.begin(LCD_COLS, LCD_ROWS);
18   if(status)
19     hd44780::fatalError(status); // does not return
20   rx.setFrequency(10180);        // frequency(in MHz) + 100
21   rx.setAgc(1);                  // enable automatic gain control
22   rx.setExtendedVolumeRange(1);  // set lower volume scale
23   rx.setVolume(10);              // set volume (range 0..15)
24   // rx.setFmDeemphasis(1);      // for the European FM standard
25   rx.setRds(true);               // enables RDS
26   rx.setRdsMode(1);
27   pinMode(seek_down_pin, INPUT_PULLDOWN);
28   pinMode(seek_up_pin, INPUT_PULLDOWN);
29 }
30
31 char str[20];
32 int i = 0;
33 void loop()
34 {
35   //remove comment signs below to wait for a button press
36   //while(!(digitalRead(seek_down_pin)||digitalRead(seek_up_pin)))
37   //   delay(5);
38
39   if (digitalRead(seek_down_pin) || digitalRead(seek_up_pin))
40   { // clear RDS data when changing station
41     rx.clearRdsBuffer();
42     lcd.clear();
43   }
```

8-6a FM-receiver sketch part 1.

```
44    if (digitalRead(seek_down_pin))
45       rx.seek(0, 0);
46    if (digitalRead(seek_up_pin))
47       rx.seek(0, 1);
48
49    delay(30);
50
51    if (rx.getRdsReady())
52    {
53       lcd.setCursor(0, 1); // 2nd row
54       lcd.print(rx.getRdsText0A());
55    }
56
57    if (i++ % 20 == 0) // every 20th time...
58    { // fetch information and display on screen
59       float f = rx.getRealFrequency() / 100.0;
60       int rssi = rx.getRssi();
61       int stereo = rx.isStereo();
62       snprintf(str, 18, "%6.2f MHz  %2d", f, rssi);
63       lcd.setCursor(0, 0); // 1st row
64       lcd.print(str);
65       if (stereo)
66          lcd.print(" S");
67       else
68          lcd.print("  ");
69    }
70 }
```

8-6b *FM-receiver sketch part 2.*

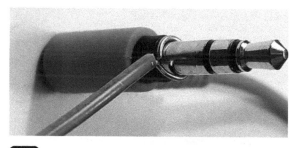

8-7 *Adding an antenna wire to your headphone plug.*

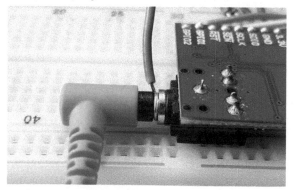

8-8 *The antenna wire should just fit between your headphone plug and the jack on the breakout board.*

measuring the signal strength. (As a rough guide to the scale, 10 is weak and 30 or above is strong.)

The headphone cable also works as an antenna with this receiver module. I found that adding 4 feet of hookup wire to the headphone ground connector improved reception. Strip half an inch of insulation from the hookup wire, bend it around the headphone plug, as shown in Figure **8-7**, and squeeze it between the headphone plug and the jack on the radio module, as shown in Figure **8-8**.

Some FM radio stations transmit the station name and other information using a digital protocol called RBDS in the US and RDS elsewhere. These abbreviations stand for "Radio Broadcast Data System" or "Radio Data System." If this data is present in the broadcast, and the signal is strong enough, the receiver chip detects it and the sketch displays the station name on the LCD screen.

If you hear a periodic noise when listening to a weak station, I suggest disabling the station-name display and only updating the display once after changing stations. To do that, look at lines 36 and 37 in the sketch. The two slashes (//) tell the compiler that any text until the end of the line should be considered a comment and ignored. I have used such comments in previous sketches to add brief explanations. If you remove the slashes from lines 36 and 37, those lines become part of the code, and the sketch will wait for one of the tuning buttons to be pressed before it proceeds to update the display. Similarly, you can add two slashes to line 57 to make it inactive, or, as it is sometimes described, to comment out the lines.

You may notice that the station name takes time

to appear or that some characters are wrong. This happens when noise causes errors in the received data. The RBDS protocol has a mechanism for error detection and correction, but it seems either the receiver chip or the Arduino library for using it doesn't implement error detection perfectly.

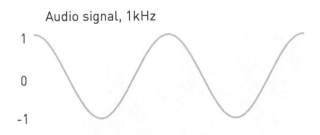

Audio signal, 1kHz

HOW IT WORKS

FREQUENCY MODULATION

Frequency modulation means that the frequency of the carrier is varied to convey information such as an audio signal. An example is shown in Figure **8-9**. (Remember that in amplitude modulation, the carrier's amplitude is varied, while the carrier frequency is constant.) The frequency mentioned for a radio station or displayed on a receiver is actually the **center frequency**, while the actual instantaneous frequency of the signal that the transmitter sends out varies around the center frequency. A common standard for broadcast stations is to allow the carrier frequency to deviate by up to 75 kHz above or below the center frequency.

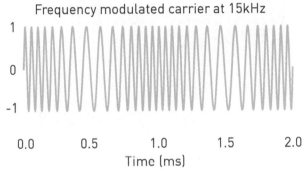

Frequency modulated carrier at 15kHz

8-9 *An audio signal (above) and the corresponding frequency-modulated carrier wave (below). The carrier frequency has been made very low and the amount of frequency modulation has been exaggerated to show it more clearly.*

STEREO SOUND

Stereo sound requires two audio channels: one for the left speaker and one for the right speaker. Stereo FM uses a clever scheme to be compatible with older mono receivers; instead of transmitting the left (L) and the right (R) channel signals, the sum *L* + *R* and the difference *L* − *R* are transmitted. Mono receivers only pick up the sum signal, while stereo receivers reconstruct the *L* and *R* signals from the two signals. This is how it's done. Adding *L* + *R* and *L* − *R* gives **2 * L**, while subtracting them gives **2 * R**, like this:

(L + R) + (L - R) = 2 * L
(L + R) - (L - R) = 2 * R

The difference signal **(L - R)** is transmitted by first amplitude-modulating

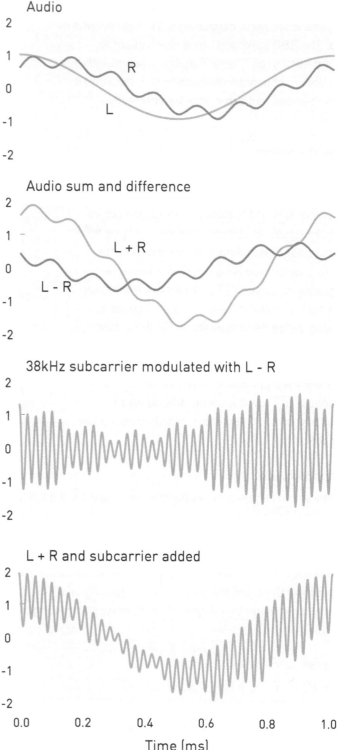

8-10 *Steps for constructing a stereo FM signal. The first graph shows the two audio channels L and R. As an example, a 1kHz tone is present in both channels. Additionally, a 7.6kHz tone is present in the R channel. In the second graph, the sum and difference of the channels are calculated. A 38kHz signal, the subcarrier, is amplitude-modulated using the L - R signal (third graph). The modulated subcarrier is added to the L + R signal (fourth graph). This signal is used to frequency-modulate the final carrier wave with a frequency in the FM broadcast band near 100MHz. (This is not shown, because the frequency is so high that the signal does not fit in the same graph, but the principle is the same as in Figure 8-9.)*

a 38kHz *subcarrier* signal with the *L* − *R* difference and then adding this modulated signal to the *L* + *R* signal. Figure **8-10** illustrates this.

This combined signal is then used to frequency-modulate the final carrier. The RDS/RBDS data is sent by modulating another subcarrier at 57kHz, again with the idea that a receiver that does not care about the data can simply ignore this frequency. In this experiment, all the processing of these signals is handled by the FM receiver chip, which uses a combination of analog-radio techniques and digital signal processing to demodulate the FM signal and recover the stereo audio signal and the data signal.

THE SKETCH

Line 4 in Figure 8-6 includes the SI470X library you installed earlier, which the sketch will use to communicate with the radio chip. Line 6 creates an object named **rx** for controlling the radio chip. The **setup** function in the sketch starts on line 14 and contains the familiar functions for initializing the screen, as well as several calls to set up the receiver chip, all of which start with **rx**.

Line 16 initializes the radio. I found that this needs to happen before the LCD screen is initialized; otherwise, the communication on the I2C bus meant for the LCD screen seems to confuse the radio chip. Line 20 sets the initial reception frequency as 100 times the frequency in MHz. (You can substitute the frequency of your favorite station here.) Line 21 turns on automatic gain control, which seems to improve the reception slightly. Line 23 sets the audio volume. The volume value is an integer in the range 0–15, with 0 being silent. Line 22 turns on something called extended volume range. On the normal volume scale, which is active by default, even the value 1 is uncomfortably loud when listening with headphones. The extended volume range is less loud, and more suitable for headphones.

Line 24 is optional because it controls a setting for an audio filter that has different standard values for FM broadcasts in different regions. The default setting of the radio chip is suitable for the US standard, and therefore I have left the line as a comment. Including line 24 will set the value to the European standard. The difference in practice is rather small but does slightly affect the balance between bass and treble in the received signal.

Line 25 turns on processing of RDS/RBDS data messages, and line 26 enables more error checking for those messages. Lines 27 and 28 set up the input pins for the two pushbuttons to enable tuning up and down. When the setup function ends, the radio chip should be receiving the frequency set on line 20. The only thing left for the loop function to do is to handle key presses on the tuning keys and display information on the screen.

Lines 39–43 check if either of the tuning buttons is pressed, and if so, it resets the RDS/RBDS data to prevent an old station name from showing.

Line 44 checks if the S2 seek-down button is pressed. If so, the **seek** command on line 45 instructs the radio chip to seek the next station at a lower frequency. Lines 46 and 47 do the same for seeking upward with the S3 seek-up button.

Line 49 is a delay for 30 milliseconds, which ensures that it's not communicating with the chip too frequently.

Line 51 checks if valid data has been received over RDS/RBDS, and if so, line 53 moves the LCD cursor to the second line. Line 54 processes the RDS/RBDS data and retrieves a string containing the name of the radio station. Lines 58–69 form a block of code for fetching information from the radio chip and displaying it on the screen. Line 57 ensures that this block is executed every twentieth pass of the loop function. Updating the display more frequently causes noticeable noise in the reception, and I found that updating the display every twentieth time reduces the noise but is enough to feel responsive.

The message displayed on the screen contains the currently tuned frequency, a number for the signal strength, and the letter **S** if the radio chip indicates that the station is received in stereo. **RSSI** in the function name stands for "received signal-strength indicator," a common name for a number indicating a signal strength.

Lines 36 and 37, which are commented out, can be used to wait for one of the tuning buttons to be pressed. That means the display is not updated while listening to a station, which I found reduces the noise when listening to weak stations. If you choose to use them, also remove line 57 (by making it a comment) so that the display is updated once per tuning.

MODIFICATIONS

Next, I'll suggest some modifications you can try to either make the radio portable by powering it with batteries or to add speakers instead of headphones. I'll also show you how to extend the frequency range for use in Japan or other regions with a different FM band.

MAKE IT PORTABLE

Powering this project from a computer through the Pico's USB port may not be ideal for practical use. You could use a USB power supply module to power the circuit or power it with batteries. The Pico can safely be powered with up to 5V through the VSYS pin (Pin 39) as long as the USB cable is not connected.

Three AA batteries in series gives 4.5V, which is also sufficient for the LCD to give decent contrast. Two AA in series gives 3V, which is enough for the Pico and the radio module but not the LCD.

The current consumption is about 65mA, dropping to 50mA if you turn off the LCD backlight by removing the jumper on the back of the I2C module. At 50mA, three alkaline AA batteries should last for about two days of continuous use. About one-half of the current goes to the microcontroller; the other half goes to the radio module. Perhaps it would be possible to put the Pico into a power-saving sleep mode most of the time since the Pico doesn't need to do much once the radio chip is set up—especially if you disable the RDS/RBDS reception and the station-name display. For a smaller radio, you could even omit the LCD and use just two AA batteries in series.

Keep in mind that the receiver is sensitive to noise from the power supply, so depending on the quality of the charger or power supply you use, it may add a significant amount of noise when receiving weak stations.

ADDING SPEAKERS

Another modification you can consider is to add proper speakers and turn the circuit into a desktop radio. I found that the amplifier on the radio module can drive a pair of small speakers, even though the volume is quite low. To connect speakers to the module, you can use an audio cable with ⅛" plugs on each end to connect the audio output to the audio jack with screw terminals you used in Experiment 3. Then, wire the speakers to the screw terminals. One speaker is wired between L and ground, and the other between R and ground. Remember the advice from Experiment 1: Speakers sound much better when mounted in boxes.

You probably want to comment out line 22 in the sketch and increase the volume on line 23. As a reminder, the maximum value is 15.

Perhaps you want to add a volume control. For that, you need two more pushbuttons—one for increasing volume and another for decreasing it—wired to any free GPIO pins on the Pico. You then have to add code to the sketch for handling these buttons, similar to how the S2 (down) and S3 (up) seek buttons are programmed. Use the functions **rx.setVolumeUp()** and **rx.setVolumeDown()** to change the volume.

When using speakers in this way, the audio cable still acts as the antenna. If the reception is poor, you can add a wire antenna to the audio connector on the radio module, as I suggested above, or add a wire antenna to the screw terminal common to both speakers.

SELECTING THE TUNING RANGE

The sketch assumes that the FM broadcast band ranges from 87.5MHz to 108MHz, which is standard in the US, Europe, and India. If you are in Japan or another country with a different FM broadcast band, you can insert the following line between lines 19 and 20:

rx.setBand(1);

for the frequency range 76MHz to 108MHz or

rx.setBand(2);

for 76MHz to 90MHz.

FINDING MORE INFORMATION

The Si4703 receiver chip is documented in a datasheet, with additional information in documents called application notes. The datasheet describes how the chip can be controlled over the I2C bus by writing values into different registers. The radio library used in this experiment is documented on GitHub at

github.com/pu2clr/SI470X

and

pu2clr.github.io/SI470X/extras/apidoc/html

These webpages provide further information about the functions the library provides.

The WRL-12938 circuit board with the Si4703 chip is described on SparkFun's product listing at

sparkfun.com/products/12938

You can also find a schematic of the radio module there, under the "Documents" tab.

STATUS UPDATE

This experiment provided you with the easiest possible introduction to FM radio and enabled you to receive transmissions by using your Pico with a little accessory board. The remaining experiments will advance from everyday transmissions of AM and FM to topics ranging from shortwave radio to metal detectors.

But first, you'll learn about remote controls that you can use to activate almost any device in your home.

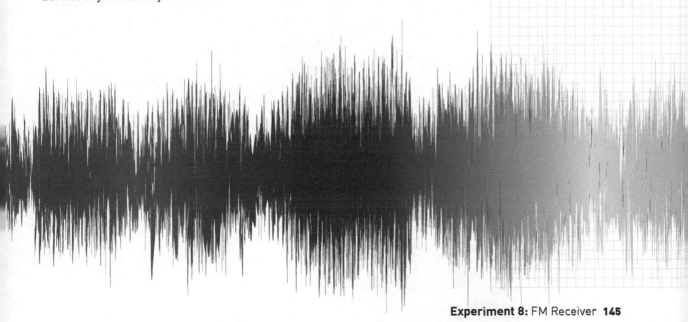

experiment

9

REMOTE
CONTROLS

The type of remote that you normally use to control a TV, a Blu-ray player, or a surround-sound system uses infrared light to send a signal. It has the big advantage of being low cost, but one obvious disadvantage: The gadgets that you control must be able to "see" the remote in your hand.

Remotes that transmit radio signals are much more versatile. A common application is to open a garage door when you approach it in your car, but there are many more situations where radio remotes work better than infrared. You often find them in doorbells, and you can program them yourself to activate lamps, audio systems, and switchable power outlets.

Off-the-shelf programmable remotes are available with multiple buttons to switch as many as 10 devices on and off throughout your home, and you'll learn about the codes that they use. Alternatively, you can add inputs to your Pico to make it do the same thing. This experiment will explain both options, and you'll see how easy it is to get a radio remote and a receiver to communicate with each other.

SENDING A SIGNAL

This experiment will require a transmitter to send a radio code and a receiver to pick it up and activate some kind of appliance. We will use a preassembled set of modules operating at

You Will Need:

- WPi469 433MHz RF preassembled Wireless Module Set (1 set, consisting of one transmitter and one receiver). Sold under brand names Whadda, Velleman, or Pimoroni. Available from sources including Jameco, RobotShop, AliExpress, eBay, and DigiKey.
- Optional: A second breadboard, Raspberry Pi Pico, and USB cable in addition to the breadboard, Pico, and cable you should already have on hand.
- Optional: A remote control operating on 433MHz with one or more buttons. Make sure it supports one of these protocols: EV1527 or PT2262. Searching for "433MHz EV1527" on eBay or Amazon works.
- Optional: A 433MHz doorbell button, supporting one of the protocols mentioned above.
- Momentary switches, with two pins, suitable for insertion in a breadboard (6).
- Resistors: 100 ohms (1), 330 ohms (1), 1K (1), 10K (1).
- Ceramic capacitors, 100nF (2).
- Electrolytic capacitors, 100µF (2).
- Generic 5mm LED (2).
- 2N3904 bipolar NPN transistor (1).
- The speaker or passive piezo buzzer from Experiment 1 (1).

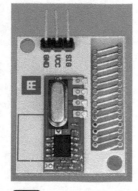

9-1 *Sample remote controls for 433MHz: a single button (doorbell), and three multibutton remotes suitable for turning lamps on and off. All work with the Pico receiver. The rightmost remote has a slider switch to select one of four groups, extending the number of devices that can be controlled.*

9-2 *The transmitter board.*

9-3 *The receiver board.*

9-4 *The receiver module inserted in a breadboard next to a Pico. Check Figure 9-5 to position the pins of each module precisely.*

433MHz, which is a commonly used frequency for this purpose. Each module is designed to be controlled by an Arduino or Pico microcontroller.

If you prefer not to buy two microcontrollers, you can use an off-the-shelf wireless-remote transmitter, such as those in Figure **9-1**. But you will still need a Pico to decode the signal.

We will start by using two Picos because this is actually the simpler option. If you only want to use one Pico, read through this first demo and make sure you understand it before continuing to the "Learning Codes of Other Remotes" section, where I will show you how to use the Pico to receive signals from ready-made remote control transmitters.

The boards by Whadda or Velleman, which I recommend, are shown in Figure **9-2** (the transmitter) and Figure **9-3** (the receiver). On each board, the gold panels with slanted lines are an antenna; you can think of them as being like coils that have been squashed flat. Turn each board over to see the back side of each coil. Because the boards are often sold without any identifying text, you may want to label them *T* and *R* so that you don't get them mixed up.

Each board has three pins that are spaced to fit your breadboard. The pin labeled *SIG* transmits or receives a signal, the pin labeled *VCC* requires a power supply of 3.3V, and *GND* is for negative ground.

Using two breadboards, place a Pico on each one and stand the transmitter module in one board while the receiver module goes in the other, as shown in Figures **9-4** and **9-5**.

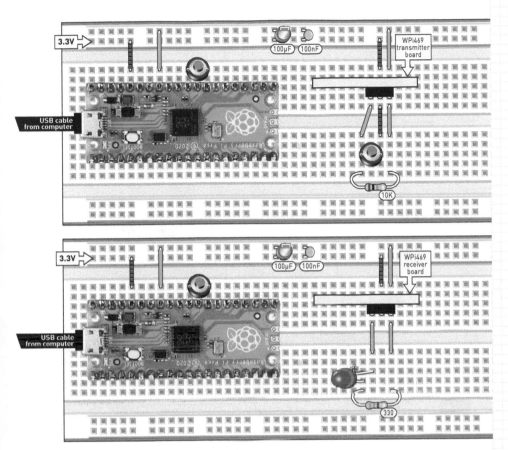

9-5 *The transmitter module (top) and receiver module (bottom) placed in separate breadboards.*

Figure 9-5 shows the wiring for the transmitter (top) and receiver (bottom). Figure **9-6**, on the following page, shows the same circuit as a schematic. The momentary switches (pushbuttons) must have two pins only, spaced 0.2" apart, so that they will fit into the breadboards as shown.

Note that this experiment does not require the LCD display that you used previously, so the wires to it have been removed. You are starting over now with new wiring around the Picos.

I am assuming that you have two USB ports available on your computer. Initially, each USB port will be used only to supply 5V. Each Pico then supplies 3.3V from Pin 36 to the transmitter or the receiver. The Picos will have more interesting tasks to perform in the near future, at which time

TRANSMITTER

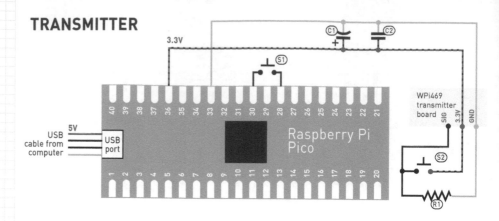

RECEIVER

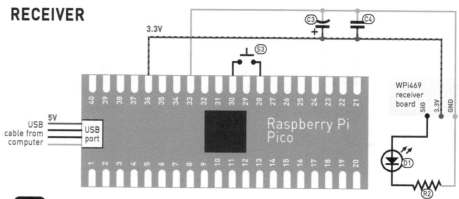

9-6 *The schematic version of Figure 9-5.*

you will have to program them separately. Right now, they do not have to be programmed with sketches, and they just draw power from the computer, so they can be plugged in simultaneously.

After you have applied power to your breadboards, observe the LED on the receiver breadboard. It may be on, off, or flickering. If you press down and release momentary switch S2 on the transmitter breadboard, the LED on the receiver breadboard should react. (Depending on your receiver, you may see different reactions, but I think you will notice a change in the LED brightness or flickering when pressing down and when releasing the pushbutton.) If you have a 433MHz radio remote control, such as any of the ones in Figure 9-1, and you press one of its buttons, the LED will flicker rapidly as it receives the coded signal.

The receiver output goes high when it detects a transmission and low when it doesn't. The receiver is built for rapid pulses (around 1kHz)

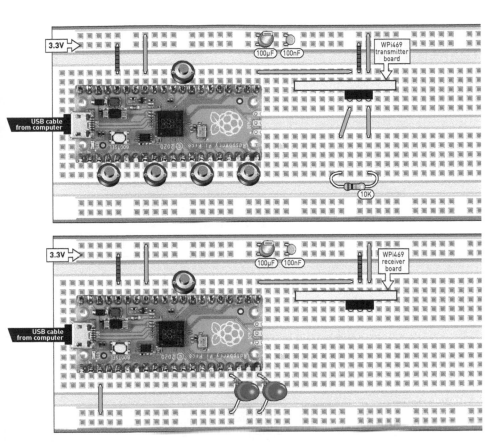

9-7 *Modifying the transmitter module (top) and receiver module (bottom) so that they are now controlled from each of the Picos.*

and automatically adjusts to the background noise level on its receiving frequency. It will not faithfully match the slow signals you can send with the pushbutton, but this test shows that the receiver reacts to the transmitter. To transmit something useful, faster signals are needed. The Pico can help you to send, receive, and decode them.

MORE BUTTONS

Figures **9-7** and **9-8** (the latter on the following page) show how you need to modify the transmitter and receiver breadboards for the next step in this experiment so that they are controlled by each of the Picos instead of just being powered by them. Notice the green jumper wire that has been added from Pin 21 of each Pico to the signal pin of each board.

In the transmitter circuit, momentary switch S2 has been removed, and switches S4 through S7 have been added. (Actually, you can take S2 from the

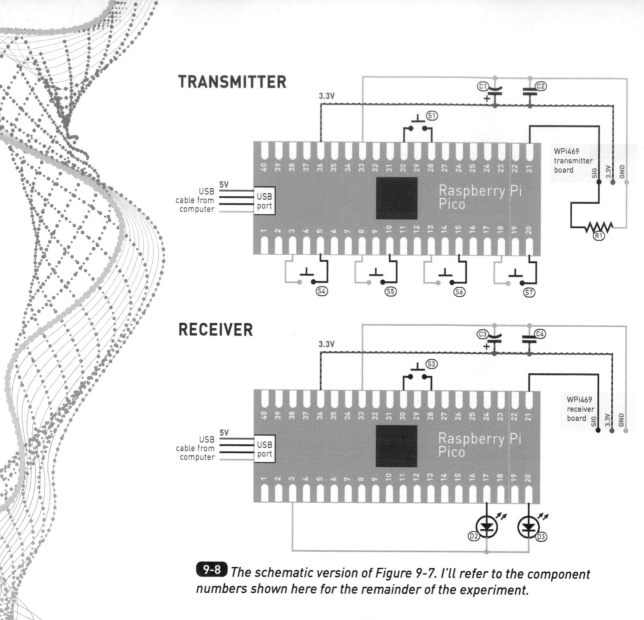

TRANSMITTER

RECEIVER

9-8 *The schematic version of Figure 9-7. I'll refer to the component numbers shown here for the remainder of the experiment.*

previous circuit and use it as S7.) In the receiver circuit, R2 and D1 have been removed, and LEDs D2 and D3 have been added. (Actually, you can reuse D1 as D2. I am renumbering the components so that I can refer to them without confusion in the text that follows.) In the receiver circuit, don't forget to add the blue jumper wire connecting Pin 3 of the Pico with the negative bus at the bottom of the breadboard. This is necessary to ground the LEDs.

Each of the buttons from S4 to S7 grounds an IO pin on the Pico when you press it. These buttons are not connected with the ground bus on the breadboard. We are using the convenient fact that some pins of the Pico are internally grounded.

In the transmitter module, we have retained R1 grounding the signal pin. If the Pico for any reason leaves the transmitter data signal floating, this 10K pulldown resistor ensures that the transmitter stays off. (Otherwise, it may transmit randomly and cause interference for other remote controls.)

Now, you will need to program each Pico with an appropriate sketch to control the transmitter-receiver modules. But before you can do this, you need to install the rc-switch library in the Arduino IDE on your computer. This library contains code to encode and decode pulse sequences used by many remote controls, making your receiver and transmitter circuits compatible with existing remote controls.

The procedure is similar to when you installed the LCD library in Experiment 4. This procedure does not require either of your Picos to be connected to your computer:

- Launch the Arduino IDE.
- Select **Tools > Manage Libraries**.
- In the search box, type **rcswitch**.
- You should see **rc-switch by sui77**.
- Click the **Install** button.

Now, plug in the Pico that controls the transmitter module, and open the **transmitter.ino** sketch in your IDE window. You should see the listing shown in Figure **9-9**. Click the Upload arrow to install it into your Pico. The sketch will remain in the memory of the Pico.

When you upload the sketch, you may see a message stating that the rc-switch library may be incompatible with your current board. I have verified that the library works on the Pico, so you can ignore this message.

```
1 #include <RCSwitch.h>
2 RCSwitch tx = RCSwitch();
3
4 const int tx_pin = 16; // pin 21
5 const int N = 4;   // number of switches
6 const int pin[]  = {       3,       7,      11,       15};
7 const int code[] = {1328149, 1328148, 1315861, 1315860};
8                   // pin  5,      10,      15,       20
9
10 void setup() {
11   tx.enableTransmit(tx_pin);
12
13   for (int i = 0; i < N; i++)
14     pinMode(pin[i], INPUT_PULLUP);
15 }
16
17 void loop() {
18   for (int i = 0; i < N; i++)
19     if (digitalRead(pin[i]) == 0)
20       tx.send(code[i], 24); // number of bits in code
21 }
```

9-9 *The sketch for your Pico that controls your transmitter module.*

```
1 #include <RCSwitch.h>
2 RCSwitch rx = RCSwitch();
3
4 const int rx_pin  = 16; // pin 21
5 const int led1_pin = 13; // pin 17
6 const int led2_pin = 15; // pin 20
7
8 void setup() {
9   Serial.begin();
10   rx.enableReceive(rx_pin);
11   pinMode(led1_pin, OUTPUT);
12   pinMode(led2_pin, OUTPUT);
13 }
14
15 void loop() {
16   int code;
17   if (rx.available()) {
18     code = rx.getReceivedValue();
19
20     Serial.print("Received ");
21     Serial.print(code);
22     Serial.print(", ");
23     Serial.print(rx.getReceivedBitlength());
24     Serial.print(" bits, protocol ");
25     Serial.println(rx.getReceivedProtocol() );
26
27     if (code == 1328149) digitalWrite(led1_pin, 1);
28     if (code == 1328148) digitalWrite(led1_pin, 0);
29     if (code == 1315861) digitalWrite(led2_pin, 1);
30     if (code == 1315860) digitalWrite(led2_pin, 0);
31
32     rx.resetAvailable();
33   }
34 }
```

9-10 *Receiver sketch.*

Unplug your transmitter Pico and plug in your receiver Pico. Open a new sketch, named **receiver.ino**, in your IDE. It should look like the listing in Figure **9-10**.

Click the Upload arrow to upload the sketch to the Pico that controls the receiver module. You can now close the IDE.

Reconnect your transmitter Pico to a USB port so that both boards are now receiving power from your computer. Press S4, and it turns on LED D2. Press S5, and it turns off the LED. Press S6, and it turns on LED D3. Press S7, and it turns off the LED.

The LEDs are just being used to verify that the system works. You can use the outputs from your receiver Pico to control almost any other device, using a transistor or a relay. Also, once you understand how the sketches work, you can modify them to control more than two devices.

Next, I will tell you more about the transmitter and receiver module and explain the sketches.

HOW IT WORKS

The modules we're using here are for the frequency 433.92MHz (although often, just 433MHz is specified). This is a frequency used for low-power, short-range remote controls. They are often sold as ASK or OOK transmitters and receivers. *ASK* stands for **amplitude-shift keying**, meaning that the signal amplitude is varied to transmit a message. That sounds similar to AM, and it is, but the term *ASK* is generally reserved for digital signals. *OOK* means **on-off keying**. With these transmitters, the two terms mean the same thing: The signal is either on or off.

These receivers and transmitters all use exactly the same frequency, so they will all hear each other if they are within range. The transmission range can be at least 100 feet—or more if there is a free line of sight between transmitter and receiver.

So, how do receivers in different devices know which signals are meant for them? And what if your neighbor also uses remote-controlled machines on the same frequency? The answer is that each remote control sends its own *digital code* in the form of a sequence of pulses. Receivers continually listen to all the signals they can find but react only to codes they recognize. What happens if two transmitters happen to transmit at the same time? There is a risk that neither message is received. Such message collisions are supposed to be rare because this type of radio device is only allowed to transmit briefly—when the user presses a button, or at preset intervals, as when a transmitter sends a temperature reading. There are limits to how often the transmission can occur and how long it can be.

THE SKETCHES

Now that you know what the transmitter and receiver modules are doing, you can understand the sketches. The general idea is that the transmitter waits until any of its buttons is pressed. Then it sends out a code corresponding to that button, using the rc-switch library you installed for the actual sending.

The receiver uses the same library to listen for valid codes. If a valid code is found and it matches one of the predefined codes, the sketch updates the state of the output pins connected to the LEDs. The receiver sketch also prints out any code received over the USB-serial port. You can read these messages using the Serial Monitor in the Arduino IDE and use this to learn the codes transmitted by other remote controls you may have.

TRANSMITTER SKETCH

Look at the transmitter sketch in Figure 9-9. Line 1 tells the compiler the sketch will use the rc-switch library (which you installed earlier), and line 2 uses that library to declare an object named **tx** (a common shorthand for *transmitter*) which we will use to interact with the transmitter module. Then follows definitions of some constants: **tx_pin** on line 4 is the pin connected to the transmitter module, and **N** on line 5 is the number of switches connected.

Lines 6 and 7 define two arrays called **pin** and **code**. The pin array contains the numbers of the pins connected to the pushbuttons, and the code array contains the codes you want the sketch to send out when those buttons are pressed.

The setup function initializes the **tx** object on line 11 and tells it which pin on the Pico the transmitter is connected to. The **for** loop on lines 13 and

14 declares all the button pins as inputs and enables the internal pullup resistors. That means these pins will read as high until the buttons pull them low by connecting them to negative ground. The loop function reads every pin, and as soon as it finds a pin with a low value, meaning that the button is pressed, it transmits the corresponding code using the function **tx.send()**. The function needs two parameters: the actual code to send and the number of bits it contains. (Twenty-four bits is a common number for remote controls. You probably don't need to change that unless you want to match an existing remote control.)

RECEIVER SKETCH

Next, look at the receiver sketch in Figure 9-10. Line 1, again, includes the rc-switch library, and line 2 declares an object to interact with the receiver. This time, it's named **rx**, for *receiver*. Lines 4 to 6 define pins for the receiver module and the two LEDs.

The setup function initializes serial communication, which actually takes place over the USB cable in this case. It will be used to write messages that can be read on the connected computer (using the Arduino IDE's Serial Monitor). Line 10 initializes the rc-switch receiver object and tells it which pin the receiver is connected to. Lines 11 and 12 set the LED pins as outputs.

The loop function checks if a new message has been received on line 17. If so, a message is printed on the Arduino serial port, on lines 20 to 25. Lines 27 to 30 compare the received code to four code values, the same ones assigned to the transmitter buttons in the transmitter sketch. If a code is recognized, one of the LED pins is pulled high or low. Finally, on line 32, the function **rx.resetAvailable()** is used to tell the library we are done processing the message and are ready for a new one.

COMMERCIAL REMOTES

In Figure 9-1, you saw some off-the-shelf 433MHz remote controls that are compatible with the receiver module wired to your Pico. You can find many more remotes online. Just search for

```
radio remote 433MHz
```

Although all the remotes that you will find share the same frequency, they use different ***protocols***—standards for how many bits (binary digits 0 and 1) are sent in a message, and how the bits are represented with transmitted signals of different length. The rc-switch library can handle many of the protocols but not all of them. So, when you're shopping for remote controls, it's best to check each part description for protocols that are known to be supported. Protocols called EV1527 and PT2262 work well. EV1527 and PT2262 are actually type numbers of IC chips, called encoders, used to generate the pulse sequence. The middle two remote controls in Figure 9-1 contain EV1527 chips.

LISTEN TO THE CODE

If you have a high-impedance earphone or a passive piezo transducer from Experiment 1, you can try the following simple experiment: Connect the earphone or beeper between the receiver module's signal pin and negative ground. Press buttons on your remotes and listen. Each remote control generates its own unique code. You may be able to hear some differences.

COMMERCIAL RECEIVERS

You can also buy an off-the-shelf receiver module that can be paired to respond to the unique code of an off-the-shelf transmitter module. To make sure that you find one containing a relay that will switch devices in your home such as lights or locks, search for

`receiver module 433MHz relay`

Figure **9-11** shows a receiver that responds to a single-button transmitter. After you pry off the plastic lid with pliers or a screwdriver, you will find a board inside with a relay, five screw terminals, a tiny momentary switch (known as the learning button), and a surface-mount LED, as shown in Figure **9-12**.

9-11 This little receiver module measures about 1" x 2" and contains a relay that can switch small devices in your home.

9-12 Inside the receiver module from Figure 9-11.

9-13 *The underside of the board shown in Figure 9-12. The screw terminals are labeled as explained in the text.*

9-14 *Programming a receiver module to recognize a transmitter. The meter is checking continuity between the relay contacts.*

The receiver needs 12VDC to operate. When you turn the board over, as shown in Figure **9-13**, you'll find that two of the terminals are labeled V+ and V-. You can use a 12V AC adapter to supply positive and negative DC power to them. The terminal labeled *NO* connects with the normally open contact inside the relay. The NC terminal connects with the normally closed contact inside the relay, and the COM terminal connects with the contact that is common to NO and NC. (That is, it connects with one of them or the other, depending whether the relay is activated.)

Figure **9-14** shows the receiver with a key chain remote and a meter that is being used to check continuity between the relay terminals. To train the receiver to recognize the transmitter, press the learning button once. The LED on the board will light up. Press the button on the remote, and the LED will flash and go dark. Now, this particular receiver will respond to this particular remote control. In the future, when you press the button on the transmitter module, the relay closes. It is safe for the relay to switch 120V so long as you use the NO and COM or NC and COM terminals.

Many receiver modules are available with a much more elaborate set of features. But I assume that you, like me, are more interested in building and programming your own.

LEARNING THE CODES OF OTHER REMOTES

Going back to your two breadboards, one with the little transmitter board and the other with the receiver board, do the following:

- Make sure that the transmitter module is *not* connected with a USB port on your computer.

- Make sure that the receiver module *is* connected with a USB port on your computer.
- Launch the Arduino IDE.
- In the menu, select **Tools > Serial Monitor**. This opens a new window, which will display any text sent by the Pico over its serial port (which is actually routed through the USB cable). The messages sent by the **Serial.print** statements in the sketch end up here.
- Press buttons on the remote, and observe the Serial Monitor window on your computer.

You should see a message—something like this:

Received 5659745, 24 bits, protocol 1

It will be repeated as long as you keep the button pressed.

Is the code of your **receiver.ino** sketch still visible in the main IDE window? If not, open it now.

You can tell your sketch to recognize the remote that you just used. Drag your mouse over the seven-digit number in the Serial Monitor window, and press Ctrl-C (Command-C on a Mac) to copy. Now, go to the sketch listing, drag your mouse to highlight code **7115876** on line 27, and paste the new code to replace it, using Ctrl-V (Command-V on a Mac).

CLONING A REMOTE

Now that you know which codes your prefabricated remote control sends, you can program the Pico on the transmitter board to mimic that remote. Take note of the codes you saw in the Serial Monitor, and replace one or more of the four codes on line 7 of the transmitter sketch. The first code on that line corresponds to button S4 on your breadboard, the second to S5, and so on.

If the "Received" message in the Serial Monitor window does not end with **protocol 1**, you will have to insert the following code as line 12 in the transmitter sketch (Figure 9-9):

tx.setProtocol(2);

where the **2** should be replaced by whatever protocol number you saw in the receiver output. This is to make sure the Pico not only sends the right code but also sends it using the same protocol as the original remote. In case you were wondering, the rc-switch library in the receiving Pico automatically selects a protocol that matches the incoming messages so you don't have to specify the protocol in the receiver.

If you now upload the modified sketch to the Pico on the transmitter board, you should be able to send the same codes as the original remote. You have cloned your remote.

Perhaps you can think of useful or fun applications for being able to clone radio remotes. Before you get too creative or worried about the security of all your remote-controlled locks, I should mention that remote controls for garage doors and wireless car keys use a different encoding scheme, even if they may use the same frequency and the same type of pulsed signaling as the remote controls I have described so far. Since it's obviously bad security to use a fixed code for such purposes, they use (or at least they really should use) something more sophisticated. One option is ***rolling codes***, a system where the transmitter generates new codes using a cryptographic algorithm, and each code is sent only once. The receiver keeps track of which codes have already been used, and will not accept old codes.

WIRELESS DOORBELL RECEIVER

Now, I will show you how to use your wireless transmitter and receiver modules to construct a wireless doorbell system. If you already have a wireless doorbell, you can add buttons or receivers to it.

In a wireless doorbell system, there are two parts: the pushbutton outside the door, which transmits a radio signal, and a receiver, which you place inside your home. The receiver creates a sound when it detects the radio signal from the pushbutton.

But what if you have a large home, and you can't hear the bell in every room? Or what if you are hearing-impaired—or you just prefer a flashing light instead of a sound?

When your Pico receiver circuit picks up the radio signal from your doorbell, you can make it create any output you choose. This can be additional to the receiver that was supplied to you with the doorbell.

For the doorbell project, you can use the transmitter breadboard you constructed earlier, or a ready-made doorbell transmitter button. I will now show you how to modify the receiver breadboard and the receiver sketch for doorbell use.

Upload the sketch in Figure 9-15 to the Pico on the receiver breadboard. Then, connect the transmitter breadboard to a USB port for power and press S4. You should see one of the LEDs on the receiver breadboard blink five times. Press S6, and you should see the other LED flash three times. For a more visible signal, you can substitute high-brightness LEDs in different colors. Having different alerts (in color or sound) can be convenient if you have different doors and is a feature I haven't seen in commercially available doorbells.

```
1  #include <RCSwitch.h>
2  const int rx_pin=16, led1_pin=13, led2_pin=15, audio_pin=14;
3  RCSwitch rx = RCSwitch();
4  void setup() {
5    Serial.begin();
6    rx.enableReceive(rx_pin);
7    pinMode(led1_pin, OUTPUT_12MA);
8    pinMode(led2_pin, OUTPUT_12MA);
9    pinMode(audio_pin, OUTPUT);
10 }
11 void loop() {
12   int code;
13   if (rx.available()) {
14     code = rx.getReceivedValue();
15     Serial.print("Received ");
16     Serial.print(code);
17     Serial.print(", ");
18     Serial.print(rx.getReceivedBitlength());
19     Serial.print(" bits, protocol ");
20     Serial.println(rx.getReceivedProtocol() );
21     if (code == 1328149) { // react to one button
22       for (int i = 0; i < 5; i++) // flash and beep 5 times
23       {
24         digitalWrite(led1_pin, 1);
25         tone(audio_pin, 440);
26         delay(250);
27         noTone(audio_pin);
28         digitalWrite(led1_pin, 0);
29         delay(250);
30       }
31     }
32     if (code == 1315861) { // react differently to another
33       for (int i = 0; i < 3; i++) // flash and beep 3 times
34       {
35         digitalWrite(led2_pin, 1);
36         tone(audio_pin, 288);
37         delay(500);
38         noTone(audio_pin);
39         digitalWrite(led2_pin, 0);
40         delay(500);
41       }
42     }
43     rx.resetAvailable();
44   }
45 }
```

9-15 *Doorbell-receiver sketch.*

To make the doorbell receiver recognize any ready-made wireless doorbell buttons that you may have, use the Arduino Serial Monitor to see which codes the buttons transmit (the sketch still prints all codes received) and adapt the sketch to recognize those codes.

Next, you can also add an audible alert. The sketch already contains the code to generate an audible square wave signal on one of the Pico's pins: the **tone(audio_pin, 288)** function call on line 36 starts outputting a square wave with a frequency of 288Hz on the specified pin, and the **noTone** function on line 38 turns it off again after a delay. All that's left to do is to connect

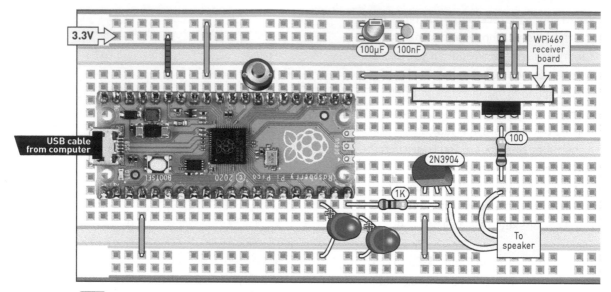

9-16 *Doorbell-receiver breadboard layout with simple audio output.*

RECEIVER

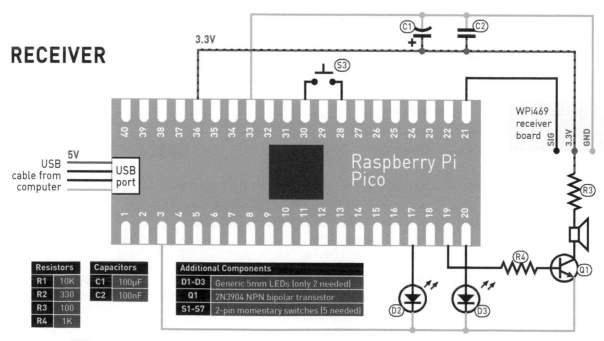

Resistors	
R1	10K
R2	330
R3	100
R4	1K

Capacitors	
C1	100µF
C2	100nF

Additional Components	
D1-D3	Generic 5mm LEDs (only 2 needed)
Q1	2N3904 NPN bipolar transistor
S1-S7	2-pin momentary switches (5 needed)

9-17 *Schematic version of the breadboard layout in Figure 9-16. The parts list is sufficient for all three circuits described in this chapter.*

a speaker so that the signal is audible. Since the signal is a square wave, we can use a simple transistor amplifier, as shown in Figures **9-16** and **9-17**.

Now, pressing the first button, S4, on the transmitter breadboard should make the receiver light up one of the LEDs and play a sound. The third button, S6, on the transmitter lights the other LED and plays another sound.

If you have the passive piezo transducer (miniature speaker) from Experiment 1, you can use that instead—just connect it between the audio output at the Pico's Pin 19 and ground. (You can omit the transistor and surrounding resistors.)

You can, of course, adapt the tone frequency and tone duration to your liking by changing the sketch, and adjust the audio volume by changing the resistor in series with the speaker. Going down to 22 ohms should work, and keeps you safely within the transistor's current limit. At this point, you may think the square wave alert sounds a bit harsh. How about some nice chimes instead? In the next section, I'll show you how to use the Pico to generate nicer sounds by using more advanced sound-synthesis techniques.

MUSICAL DOORBELL

If you want a nicer audio output, you can use the Pico to create more advanced audio waveforms.

Upload the sketch in Figures **9-18** and **9-19** (the latter on the following page) to your receiver Pico, and try the buttons on the transmitter breadboard (or your doorbell transmitter buttons) again. You should hear a sequence of tones: three tones for one button and two for the other. The tones are smoother than before and vary in intensity, rather than starting and stopping abruptly.

```
1  #include <RCSwitch.h>
2  #include <PWMAudio.h>
3  const int rx_pin=16,led1_pin=13,led2_pin=15,audio_pin=14;
4  const int RATE = 44000; // PWM frequency & sample rate Hz
5  // ADSR:   Attack rate  Decay rate  Sustain Release rate
6  float      A=10.0/RATE, D=5.0/RATE, S=0.7,  R=10.0/RATE;
7  RCSwitch rx = RCSwitch();
8  PWMAudio audio = PWMAudio(audio_pin);
9  // musical note frequencies in Hz
10 #define C4   261.6256 // C in octave 4
11 #define C4s  277.1826 // C sharp
12 #define D4   293.6648 // etc
13 #define D4s  311.1270
14 #define E4   329.6276
15 #define F4   349.2282
16 #define F4s  369.9944
17 #define G4   391.9954
18 #define G4s  415.3047
19 #define A4   440.0000
20 #define A4s  466.1638
21 #define B4   493.8833
22 void setup() {
23   Serial.begin();
24   rx.enableReceive(rx_pin);
25   pinMode(led1_pin, OUTPUT);
26   pinMode(led2_pin, OUTPUT);
27   audio.setBuffers(4, 32);
28   audio.setFrequency(RATE);
29   audio.begin();
30 }
31
32 void play(float f, float len) {
33   int16_t s;
34   float e = 0;
35   char state = 'A';
36   int remaining = len*RATE;
37   for(int i = 0; ; i++) {
38     remaining--;
39     switch(state) {
40       case 'A':
41         e += A;
42         if (e >= 1)  {
43           state = 'D';
44           e  = 1.0;
45         }
46         break;
47       case 'D':
```

9-18 *Doorbell-receiver sketch with nicer audio.*

```
48        e -= D;
49        if (e <= S)
50          state = 'S';
51        break;
52      case 'S':
53        if (remaining <= 0)
54          state = 'R';
55        break;
56      case 'R':
57        e -= R;
58        if (e <= 0)
59          return; // end of the note
60        break;
61      }
62      s = 32767 * e * sinf(2*M_PI/RATE*f*i);
63      audio.write(s);
64    }
65  }
66  void loop() {
67    int code;
68    for (int i = 0; i < 4*32; i++)
69      audio.write(0); // clear buffer - silence
70    if (rx.available()) {
71      code = rx.getReceivedValue();
72      Serial.print("Received ");
73      Serial.print(code);
74      Serial.print(", ");
75      Serial.print(rx.getReceivedBitlength());
76      Serial.print(" bits, protocol ");
77      Serial.println(rx.getReceivedProtocol() );
78      if (code == 1328149) { // react to one button
79        digitalWrite(led1_pin, 1);
80        play(C4, 0.4);
81        play(E4, 0.4);
82        play(G4, 0.8);
83        digitalWrite(led1_pin, 0);
84      }
85      if (code == 1315861)  { //  another button
86        digitalWrite(led2_pin, 1);
87        play(E4*2, 0.4);
88        play(C4*2, 0.8);
89        digitalWrite(led2_pin, 0);
90      }
91      rx.resetAvailable();
92    }
93  }
```

9-19 *Doorbell-receiver sketch with nicer audio (continued).*

I will explain how the sketch works, and how you can modify it to play your own notes.

HOW IT WORKS
SOUND SYNTHESIS

In Experiment 5, I mentioned that pulse-width modulation can be used to generate an analog output signal with a digital output. The Arduino-Pico library you are using contains functions to help play sounds using pulse-width modulation. You have to supply *samples* as a number representing the audio signal at different points in time. The library handles the work of playing those samples at a fixed rate you can choose.

Using pulse-width modulation to output audio means that the output is a square wave (with a much higher frequency than the audio signal itself). This square wave can be amplified by the simple transistor amplifier in Figure 9-16. The speaker itself acts as a filter, removing the high PWM frequency so that only the audio frequencies remain. (Besides, we chose the PWM frequency outside the audible range so that you cannot hear it anyway.)

THE SKETCH

The following is a detailed explanation of the parts of the doorbell sketch. The remote control receiving parts are the same as in the previous sketches, while the sound generation is new. Line 2 tells the compiler that the sketch will use the audio functionality. Line 3 defines pins for different purposes. The constant **RATE** on line 4 is the *sample rate* in hertz, which is the number of samples played per second. At a minimum, the sample rate must be twice as high as the highest frequency you want to play back, but a much higher rate helps to increase the quality. Another issue is that the PWM frequency is also set to this value, and for that reason, it's best to choose a value well outside the audible range.

Line 6 defines constants describing the sound, which I will explain below. Line 8 initializes the audio system, and lines 10–21 define frequencies for different notes in hertz. In the setup function, line 27 tells the audio system about the size and number of **buffers** needed. A buffer is a space for samples in the microcontroller's memory. When the program produces samples, they are stored in one buffer, while the sound system plays back samples from another. All of this is managed by the sound system, and we do not need to handle the details. Line 27 requests four buffers of 32 samples each. Line 28 sets the sample rate, and line 29 starts the audio system.

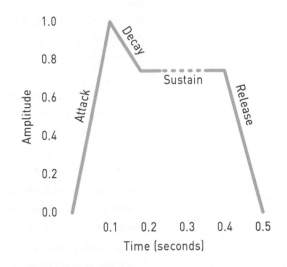

9-20 *ADSR envelope.*

The function **play(f, len)**, defined on line 33, plays a tone with frequency *f* for a duration of **len** seconds. The tone is a sine wave, but to make it more interesting, the amplitude is varied during the duration of the tone. How the amplitude varies over time is called the **envelope**. In the function, it is described with four numbers, called **attack**, **decay**, **sustain**, **release**. This type of envelope is called ADSR and is very common in synthesizers, perhaps because it's reasonably simple while leaving a lot of room for making different sounds. For example, the Commodore 64 (which I used to program a long time ago) used ADSR envelopes to shape the sound it created with its SID sound chip. (If you are curious, or old enough to become nostalgic, search online for SID music and you should find videos where you can both hear the music and see the waveforms on an oscilloscope-like display).

The ADSR envelope has four phases, shown in Figure **9-20**. During the attack phase, the amplitude rises from 0 to the maximum value 1. The attack value specifies the rate of increase—in other words, how long this phase will take. Then the decay phase starts, in which the amplitude decreases at a rate set by the decay value. That phase stops when the amplitude reaches the sustain level, where it is held during the sustain phase. The sustain phase

lasts as long as a key is held down (on a synth with a keyboard) or as long as the note is specified to be played. When the key is released or the note ends, the release phase starts, and the amplitude drops to 0 with a rate specified by the release value.

The variable **remaining**, defined on line 36, keeps track of how many samples are left until the sustain phase ends. The variable **s** on line 33 holds one audio sample in the form of a 16-bit integer (which is what the audio system expects), and the variable **e** on line 34 is the envelope value. The **for** loop beginning on line 37 produces the audio samples. Lines 39–61 are the envelope generator. This handles the four phases using a **switch** statement. The variable **state** holds one of the letters *A*, *D*, *S*, or *R* to indicate the current phase. The **switch** statement makes the microcontroller jump to the **case** statement matching the variable state. For each of the cases, the **e** value is updated, and there is a test to determine if the next phase of the envelope should start. The **break** statement on line 46 indicates that the handling of a case is finished and control should be transferred to the end of the switch block. Omitting a break is a common mistake when writing switch statements. When that happens, the microcontroller will happily continue executing the code for the next case.

Line 62 computes a sample value by multiplying the **sinf** function for a sine, the envelope to adjust the amplitude, and the constant 32,768 to scale the value to fit in a 16-bit signed integer. You can think of this as an example of amplitude modulation: The envelope modulates the amplitude of the sine function.

Line 63 plays the sample—or places it in the audio buffer to be played slightly later. The function **audio.write()** is *blocking*, meaning that if the audio buffers are full, the function will wait until there is space for the sample. That means that the play function does not need to care about timing, only about producing samples fast enough to keep the buffers from becoming empty (which would result in an audible glitch). This is an application where the high speed of the Pico is convenient—the Pico is fast enough to calculate a sine and the envelope for each sample. On a slower microcontroller, we would probably have had to read the sine values from a table in memory and would have also needed to avoid using float variables, because they slow down calculation.

The loop function is similar to the one in the previous sketch. It detects two remote control codes, and for each of them, it lights an LED, plays two or three notes, and then turns the LED off.

THE MUSICAL NOTES

The three notes for the first button form a pleasant-sounding major chord. The two notes for the second button sound like a classic ding-dong doorbell sound. Multiplying the frequency constants by 2 in the second block makes the notes one *octave* higher. These are just examples—you can modify the notes, their durations (by editing the **play** commands at lines 80–82 and 87–88), and the ADSR values on line 6 as you wish. Here is one suggested alternative for the values on line 6 of the sketch, which gives a more percussive sound:

```
float A = 50.0/RATE, D = 5.0/RATE, S = 0.3, R = 10.0/RATE;
```

Those divisions by the constant **RATE** are there to convert rise rates from how far the signal rises per second to how far it rises per sample, which is what the envelope generator expects. Remember that **A**, **D**, and **R** are amplitude change rates, while **S** is the level where the envelope is kept during the sustain phase.

If you like synth ideas, you can skip the doorbell part and just use the Pico for making music and sounds. There are lots of possibilities beyond the scope of this radio book.

OTHER MODULES AND ADDING WIRE ANTENNAS

All remote control modules require antennas. The WPI469 modules I recommended have antennas printed on their circuit boards, as shown in Figures 9-2 and 9-3. Other remote control modules require you to solder on an antenna. If you're prepared to do that, there are many other types of modules available that you can use instead.

Such remote control modules are often sold in pairs, with a transmitter and a receiver, and sometimes also include antennas in the form of helical wire coils. Transmitters and receivers from different pairs are generally compatible, as the on-off keying standard is very simple. Some models I have tested are the STX882 transmitter and the SRX882 receiver, the WL102 transmitter and the WL101 receiver, and the SYN115 transmitter and the SYN480R receiver. All of these work for the exercises in this experiment if you solder on antennas, as I will explain below. There is, however, one type of module to avoid, often described as a **supergenerative receiver**. It is cheap and widely available but much less sensitive than the other types mentioned here.

The remote control modules I recommend here use the frequency 433.92 MHz. (Even though the modules are described as just 433MHz, they are really operating at 433.92MHz.) This frequency is used globally for remote controlling purposes. Similar modules exist for 315Mhz, which is another frequency used in the United States. The 315MHz modules work in the same way but are not compatible with 433.92MHz modules.

With some modules, helical antennas made of stiff wire are included, and you just need to solder them to the correct point on the circuit board. If the modules you purchase don't include antennas, you can make your own from pieces of hookup wire of the correct length, which in this case is 6¾". Cut this length of hookup wire (precisely), strip a short section of the insulation from one end, and solder the bare end of the wire to the solder pad on the module marked *ANT* or something similar. On most modules, the antenna solder pad is separate from the pads for supply voltage and the logic input or output.

You need to solder an antenna to both the transmitter and the receiver. The antenna should be kept as free as possible from other circuitry, wires, or conductive objects. In principle, it should be straight, but if you want to fit the circuit in an enclosure, you can try bending it to fit. It will probably work, as long as the enclosure isn't metallic!

Why this length? In order to be efficient, an antenna needs to resonate with the frequency it is supposed to transmit or receive. A good length is one-quarter of the wavelength when the signal is fed into the wire from one end and the other end is unconnected. Such an antenna is called a **quarter-wave antenna**. To find out how long the antenna should be, you first need to know the wavelength. The wavelength l is the speed of light divided by the frequency. (I showed a similar relation for sound waves in Experiment 1.)

$$l = c \ / \ f$$

(l is in meters, c is 300,000,000 meters per second, and f is in hertz.)

```
300,000,000 / 433,920,000 = 0.69 meters
```

Then, one-quarter of the wavelength is 17.2cm, or 6¾".

I did not emphasize the antenna length before, when dealing with AM radio in the medium-wave band, because there, the wavelength is impractically long—160–500 meters—and we quickly moved to using ferrite-rod antennas. For the remote controls in this experiment, the wavelength is short enough to have a practical quarter-wave antenna, so I'm telling you now.

STATUS UPDATE

In this experiment, you not only learned how to make the Pico serve as a wireless remote but also saw how it can generate pleasing audio tones when serving as a doorbell.

The next experiment introduces a different type of radio receiver. Instead of being limited to commercial AM radio stations, it takes the first step toward opening up the world of shortwave radio, where licensed amateurs are free to transmit their own signals over hundreds of miles—or even further, if the conditions are favorable.

10

REGENERATIVE AM RECEIVER

In this experiment, you will build a different type of AM receiver, using just an LM386 audio amplifier chip in a creative way. The chip wasn't designed for this purpose but provides further proof that the world around us is full of radio waves.

A man named Martyn McKinney discovered the previously unknown capability of an LM386 when he was repairing a conventional radio. He found that the chip could amplify and demodulate an AM signal, even when the other components in the radio were disconnected.

It's an example of a *regenerative receiver*, which uses positive feedback to amplify the radio signal. This used to be a well-known technique back in the days of vacuum tubes, but dealing with it is an adventure, as the feedback easily gets out of control, with unpredictable results. You'll have to adjust it carefully to find the sweet spot where a suitable antenna can bring in distant transmissions.

You Will Need:

- Slide switch, SPST or SPDT, to fit breadboard (1).
- Resistors: 100 ohms (1), 4.7K (1).
- Ceramic capacitor, 220pF (1).
- Electrolytic capacitors: 10µF (2), 100µF (2).
- Inductor, 1mH (1). It should have a brown-black-red-gold (or silver) color code.
- Tuning capacitor, 200pF, type 223P, as in previous experiments (1).
- Ferrite rod, 3/8" diameter, 6" long, as in previous experiments (1).
- Trimmer potentiometer, 10K (1). A larger potentiometer is nicer to handle but won't fit on the breadboard.
- LM386 amplifier chip (1), made by Texas Instruments or National Semiconductor (now also Texas Instruments).
- Wired headphones, as used typically with a music player, with a 1/8" audio plug (1). The high-impedance earphone from Experiment 1 will also work.
- 1/8" audio plug to screw terminal adapter (1).
- Optional: Speaker, as in previous experiments (1).

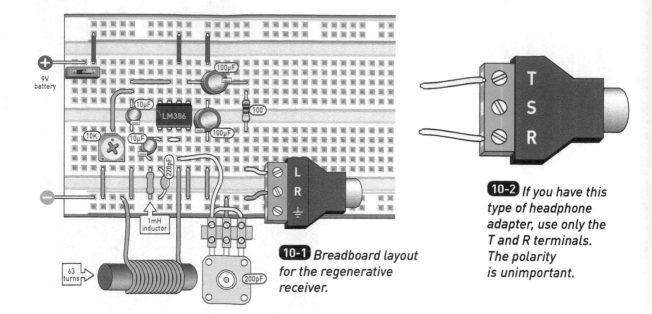

10-1 Breadboard layout for the regenerative receiver.

10-2 If you have this type of headphone adapter, use only the T and R terminals. The polarity is unimportant.

A REGENERATIVE CIRCUIT

The circuit shown in Figure **10-1** is all you need for a regenerative radio receiver.

The headphone adapter is wired in an unusual way, so that if you are using two headphones that normally work in stereo, the mono signal from this circuit passes through them in series to double their resistance. You can also use the high-impedance headphone of the type that I introduced in Experiment 1. The adapter will work either way.

Figure **10-2** shows how you should wire an adapter with terminals lettered *T*, *S*, and *R* (for *tip*, *sleeve*, and *ring*) instead of *L*, *R*, and the ground symbol (for *left*, *right*, and "negative ground").

The behavior of the circuit tends to give you all or nothing: not enough sound, or too much. It can also create some unpleasant sound effects. You need to adjust the amplification with the trimmer for each station you receive. But it's remarkably simple compared with other regenerative circuits. The schematic is shown in Figure **10-3**.

C6 is the same variable capacitor that was introduced in Experiment 1. Make sure to connect its middle pin to the negative bus of the breadboard,

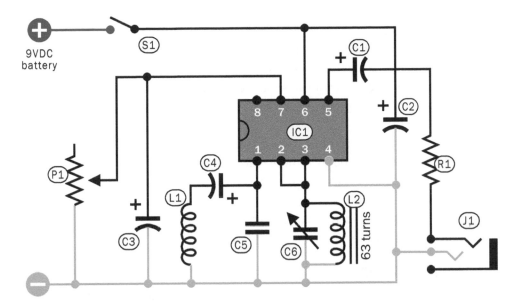

10-3 *Schematic of the regenerative receiver.*

as shown in Figure 10-1. If you fail to do this, you may find that touching the metal screw in the center of the tuning dial will create an unpleasantly loud noise in the headphone.

L2 is a coil on a ferrite rod, the same as in Experiment 2. It consists of 63 turns of hookup wire, in two layers. Check back to Figures 2-1, 2-2, and 2-3 for the method of winding it. (You will not need the second, shorter coil that was added in Experiment 2.)

When you have assembled the circuit, turn the trimmer fully counterclockwise. After switching on power from a 9V battery, start turning the trimmer clockwise, and you should hear some static in the earphones.

To find a radio station, you will need to rotate two controls simultaneously. The variable capacitor sets the tuning frequency, while the trimmer potentiometer sets the gain, or the sound volume.

If the trimmer is set too low, you won't hear anything. If you find a radio station, you will find that turning the trimmer clockwise increases the audio volume until eventually, the positive feedback gets out of hand and creates a whistling sound. This is the problem with a regenerative receiver: You have to get it just right. Decrease the feedback by turning the trimmer counterclockwise until the signal is good again.

The advice about interference in Experiment 2 still applies here. You will have to switch off all appliances and devices that generate radio interference, and you may even have to go outside to get a good signal. Also, remember that the ferrite antenna is directional—it is most sensitive to signals coming in perpendicular to the rod. Try rotating it for the best signal.

You can use a speaker (the same one as in previous experiments) instead of the headphone. To do so, remove resistor R1 and the headphone adapter J1 and connect the speaker in their place.

I found that this receiver had a comparable sensitivity to AMR2, or is perhaps even a bit more sensitive.

HOW IT WORKS

In this experiment, we are taking advantage of an **undocumented feature** in the LM386—a capability that the manufacturer didn't intend us to use. Seeing precisely how it works requires looking at the internal schematic of the amplifier, shown in Figure **10-4**. When you connect capacitor C5 from Pin 1 to negative ground, it increases the amplification to a factor of thousands, to the point where it's sensitive enough to amplify radio signals. In normal use, this pin is either left unconnected for a gain of 20 or connected to Pin 8 through a capacitor for a gain of 200 (which is what I suggested when we used the LM386 to amplify sound in Experiment 2). Here, Pin 1 is connected to ground (through an inductor and a capacitor), disabling the negative feedback entirely for audio frequencies and causing a very high gain for these frequencies.

10-4 *The interior components of an LM386 amplifier chip.*

In this receiver, like in AMR2, L2 acts as an antenna. It picks up the varying magnetic field from incoming radio waves. The variable capacitor C6 and L2 together form a resonance circuit; you select the reception frequency by adjusting the capacitance of C6. The resonance circuit is connected to the amplifier input pins 2 and 3.

Now, you may ask, how does the LM386 demodulate the signal? Just amplifying an AM signal is not enough to recover the audio. Honestly, I don't know the complete answer. There has to be an effect like that of the diode in the "crystal" radio in Experiment 1, recovering the audio by rectifying the radio-frequency signal. The base-emitter junction of the input transistors (the ones with their bases connected to the input Pins 2 and 3) can perform this function—the base-emitter junction in a transistor acts like a diode.

The trimmer in our circuit controls the voltage on Pin 7, which is normally used just to connect a voltage-stabilizing capacitor. In this circuit, the voltage on Pin 7 controls the current through the input transistors and therefore their amplification (similar to how the base current controlled the amplitude of oscillation in AMT1 in Experiment 3). This is how the trimmer controls the amount of amplification in the circuit—again, in an unofficial, undocumented way.

Anytime you use an undocumented feature of an electronic component, you take a chance on it not working—or working differently, depending on the manufacturer. Chips from National Semiconductor (as shown in the top image in Figure **10-5**) work the best in this circuit, but because that company is now owned by Texas Instruments, you may see the TI logo printed on them, and they may be described as sold by TI. Just to make things more complicated, there are three variants of the Texas Instruments LM386: LM386-1, LM386-3, and LM386-4. Each of them has slightly different characteristics, but in my tests, they performed similarly in this circuit.

10-5 *LM386 amplifiers from two different manufacturers: National Semiconductor (top) and an unknown source (bottom). Note the National Semiconductor logo, an N with curved tips, on the top.*

SHORTWAVE

Your regenerative receiver has another undocumented feature: It can be adapted to receive **shortwave** radio signals in the frequency range between 3MHz and 30MHz. These frequencies have a remarkable property: When they radiate outward from the Earth, they can reflect from a layer of the atmosphere known as the *ionosphere*. At this level, solar radiation can ionize some molecules in the air (mainly oxygen and nitrogen), meaning that they lose one or more electrons. This can enable radio waves to bounce back so that a signal can be received beyond the horizon.

Shortwave is fun because it's less predictable than medium-wave. You can hear broadcasts from far away, even the other side of the planet, because the signal can bounce between the ionosphere and the ground multiple times.

You may be able to hear radio amateurs. It is also possible you hear nothing—because the atmospheric conditions happen to be unfavorable or because the signal is drowned out in noise from local sources. The propagation of shortwave signals will vary because the properties of the ionosphere vary during the day due to the amount of incoming sunlight changes and according to how solar activity such as a solar flare affects the amount of incoming ionizing radiation. Broadly speaking, nighttime tends to be favorable for longer transmission. However, this is quite random and depends on the frequency.

Figures **10-6** and **10-7** show the circuit modified for shortwave reception.

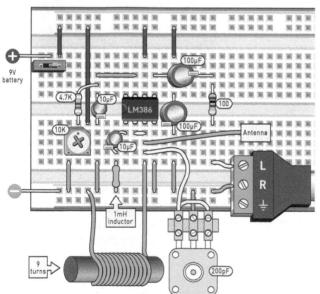

10-6 *The LM386-receiver adapter for the shortwave band.*

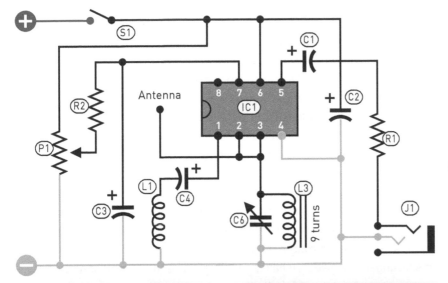

10-7 *Schematic of the circuit in Figure 10-6. L2, the 63-turn coil, was used only in the medium-wave version of the circuit.*

Resistors	
R1	100
R2	4.7K

Capacitors	
C1	100µF
C2	100µF
C3	10µF
C4	10µF
C5	220pF
C6	200pF

Additional Components	
IC1	LM386 amplifier
P1	10K trimmer potentiometer
L1	1mH inductor
L2	63 turns around ferrite rod
L3	9 turns around ferrite rod
S1	SPST or SPDT switch
J1	1/8" jack adapter

These are the changes that you need to make to the previous version of the circuit:

- Remove capacitor C5.
- Connect the remaining pin of trimmer potentiometer P1 to the positive bus.
- Insert 4.7K resistor R2 between the wiper of P1 and Pin 7 of IC1.
- Replace the 63 turns of wire on coil L2 with 9 turns. If the original coil is wound on a card-stock tube, you can try sliding it off the ferrite rod so that you can put it back later. If sliding doesn't work, just wind the new coil next to the original one.
- Connect a 9-foot length of hookup wire as an antenna to Pins 2 and 3 of IC1.

The 9-turn coil on the ferrite bar, together with the variable capacitor, sets the tuning range at the lower part of the shortwave band. However, the antenna wire adds some capacitance, which lowers the frequency by an unknown amount. While listening, you can slide the coil toward the end of the ferrite rod to decrease the inductance and increase the frequency. You can experiment with adding a ground wire connected to the negative bus on the breadboard, as in Experiment 1. This should increase the signal strength but may add a considerable amount of noise.

There is an alternative way to connect the antenna: Wind the end of the antenna wire a few turns around the ferrite bar next to the coil. Connect the short end of the antenna wire to another wire—to ground, or stretch it out for some distance roughly opposite to the direction of the first antenna. This arrangement gives a weaker coupling between antenna and resonance circuit, is supposed to work better, according to some sources, and will not change the resonance frequency much.

I have three suggestions in case you hear only noise:
• Slide the coil toward the end of the ferrite rod, or remove one or a few turns, to change the frequency range.
• Try another time, perhaps at night.
• Try a different location that is outdoors and far away from possible noise sources.

If you are lucky, you may hear Morse code transmissions as a series of short and long beeps. Morse code is nowadays used mostly by radio amateurs. In the shortwave, band there are several frequency ranges reserved for amateur-radio use, and this receiver is designed for 3.5MHz, the lowest of the shortwave bands.

Morse code transmissions (sometimes called CW, for **continuous wave**, among radio amateurs) are created by switching a carrier wave on and off—the simplest possible modulation. Each letter of the alphabet is identified by its own set of long and short pulses (represented by dashes and dots), separated from the next set by a longer pause.

On a regular AM receiver, you cannot receive Morse code. But a regenerative receiver can receive Morse code when the feedback control is tuned above the point of oscillation. When the receiver oscillates at a frequency near that of a Morse code transmission, the two signals combine to form audible beeps.

You will learn more about this mixing of signals in the next experiment, where I will show you how to use it in a metal detector, and later in another type of receiver for the shortwave band. We will also return to the topic of amateur-radio in the final chapter.

FURTHER READING

For Martyn McKinney's own description of the circuit, and many more variants, see

edn.com/create-radio-receiver-circuits-with-the-lm386-audio-amplifier

STATUS UPDATE

In this experiment, you learned about the powers and peculiarities of positive feedback, delivered by a chip that was never designed for this precise purpose. You also learned how it can receive transmissions in Morse code.

The next experiment will stay closer to home, as it shows you how to build a simple circuit that resonates with its environment, enabling you to find metal objects that may be invisible to the naked eye. This will be a step toward an unexpected conclusion: the mixing of frequencies, which will enable you to demodulate radio signals.

11

METAL DETECTOR

It's easy to build a circuit that responds to metal in the environment, and this experiment will show you how. Initially, you'll find that the circuit reacts when you place an object such as a teaspoon in the center of a large-diameter coil of wire. After completing this test successfully, you'll be able to modify the coil so that it can detect coins in someone's pocket. This demonstrates how a coil generates a magnetic field that creates electric currents in other objects, even if they are not magnetic.

You'll learn more about inductance and oscillators, and you'll discover how two high-frequency signals can combine to create a much-lower frequency within the range of human hearing. The parts list includes a chip containing logic gates, a handful of capacitors and resistors, an inductor, and some wire.

You Will Need:
- Slide switch, SPST or SPDT, to fit breadboard (1).
- Resistors: 100 ohms (2), 4.7K (1).
- Inductor, 22µH (1), Bourns 78F220J-RC or similar.
- Ceramic capacitors: 220pF (1), 470pF (4), 100nF (1).
- Electrolytic capacitor, 100µF (1).
- Variable capacitor, 200pF, type 223P (1)—the same as used in previous experiments.
- High-impedance earphone or stereo earphones (1, or 1 pair).
- 1/8" audio jack with screw terminals (1)—the same as used in previous experiments.
- 4030B or 4070B Quad XOR logic chip, or 4011B quad NAND logic chip (1).
- 22-gauge hookup wire for sensing coil (13 feet).
- Optional: Frequency-counter circuit from Experiment 7 (1).
- Optional: Nonconductive base on which to mount the sensing coil (1).
- Optional: Ferrite rod, as used in previous experiments (1).

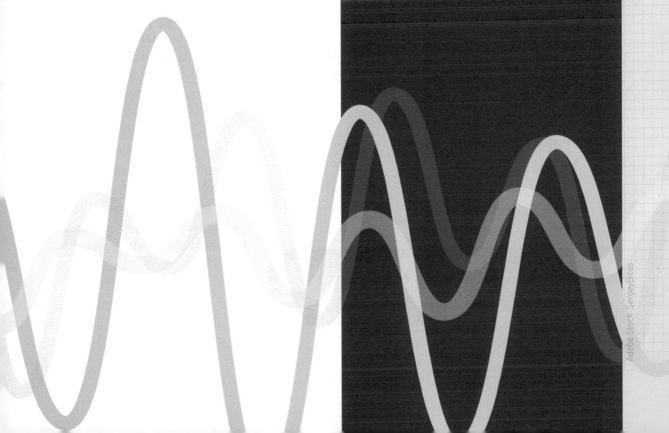

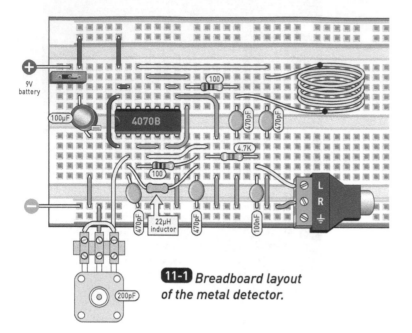

11-1 *Breadboard layout of the metal detector.*

TWO OSCILLATORS AND A MIXER

Start by assembling the circuit in Figure **11-1**. The tuning capacitor, identified as C5 in Figure **11-2**, is the same as the one that you have used previously. The audio jack output is wired in the same way as in Experiment 10, so that it will work either with conventional stereo headphones or the high-impedance earphone from earlier experiments.

ADDING A SENSING COIL

The yellow "sensing coil" in each figure will function like an antenna to make your device sensitive to nearby conductive objects. In Figure **11-3**, the large coil is 4¼" in diameter, created by wrapping 10 turns of 22-gauge hookup wire around a 2-liter soda bottle. The smaller coil contains 17 turns, 2" in diameter. Try each of them, one at a time.

The coils don't have to be absolutely, precisely circular, but you will be able to control their shape and handle them more easily if you tape each one to a piece of cardboard or plywood. Any flat surface will do, so long as it does not conduct electricity.

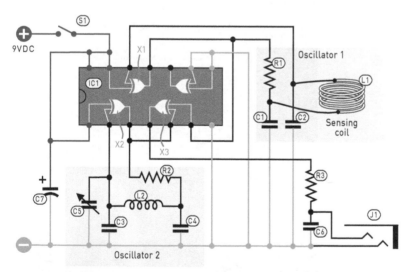

Resistors	
R1 & R2	100
R3	4.7K

Capacitors	
C1–C4	470pF
C5	200pF
C6	100nF
C7	100µF

Additional Components	
L1	10 turns of 22-gauge hookup wire
L2	22µH inductor
IC1	4070B quad XOR chip
S1	SPDT or SPST slide switch
J1	1/8" jack adapter

11-2 *Schematic of the metal detector.*

SEEKING A SIGNAL

After you connect a coil and attach a 9V battery to the circuit, turn the variable capacitor until you hear a sudden whistling sound. Continue turning the capacitor very slowly until the sound goes lower in pitch and finally disappears. Keep turning, and it should reappear and rise in pitch. Back up and set the capacitor at the silent spot between the falling and rising frequencies.

Now, move a metal object over or into the coil. If the object is similar in size to the coil, the circuit will detect the object by whistling in response.

What if you hear nothing at all? Try adding a turn of wire to your coil, or subtract a turn.

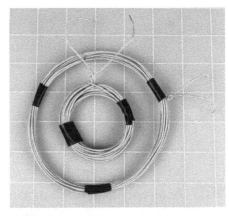

11-3 *The 2" coil contains 17 turns of wire. The 4¼" coil contains 10 turns.*

If you move your fingers toward and away from the circuit (especially C2 or the tuning capacitor), you should hear the frequency changing. The circuit is sensing objects near it that conduct electricity—including you! Your body is not a good conductor (it has a relatively high electrical resistance), but electrons can travel through the moisture in your tissues. Consequently, your body has capacitance, which affects the capacitance in the oscillator circuits.

Now, approach the sensing coil with a metal object (for example, a pair of pliers or the metal lid of a cooking pot). You should hear a tone in the earphones that increases in pitch as you move the metal object closer.

You can test which objects it can detect, and at what distance. Examples include a small coin at 2", a ferrite rod (from Experiment 1) at 5", small pliers at 2¾", or a loop of hookup wire 3" in diameter, with the ends stripped and held together, at a 5" distance. Also, try the same length of hookup wire not forming a loop—you will find it is not detected.

HOW IT WORKS

When the circuit applies voltage to the sensing coil, it can induce tiny amounts of current in any nearby object that conducts electricity, and this changes the frequency of the oscillator circuit containing the coil. Because metal objects are more conductive than people (generally speaking), the coil will react more strongly if it senses some metal nearby.

Inputs

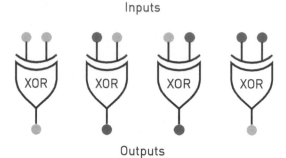

Outputs

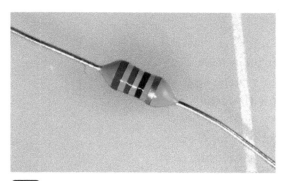

11-4 *The outputs of an XOR logic gate in response to low-voltage inputs (blue dots) and supply-voltage inputs (red dots).*

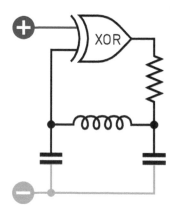

11-5 *The inductor contains a very small coil.*

11-6 *The basic principle of each oscillator.*

The only problem with using this effect is that if we choose a variable capacitor and a variable inductor that are conveniently and affordably small, the circuit will oscillate at a frequency that is much too high to hear. We can get around this limitation by using a reference oscillator in addition to the sensing oscillator and mixing their frequencies in such a way that the difference between them is at an audible frequency.

This circuit uses XOR logic gates to make this happen. In case you are unfamiliar with the XOR, its output is low when both inputs are the same and high when the inputs are not the same, as shown in Figure **11-4**, where blue dots indicate 0V and red dots indicate the supply voltage.

In Figure 11-2, an X-ray view shows you that there are four XOR logic gates inside the 4070B integrated circuit chip, identified as IC1. Oscillator 1 consists of two capacitors, a resistor, and your sensing coil, connected with the XOR gate labeled X1. Oscillator 2 consists of two capacitors, an additional variable capacitor, a resistor, and a fixed-value coil (properly known as an inductor) in a package that looks quite like a resistor. Figure **11-5** shows the inductor.

Each oscillator works by using an XOR gate as an inverting amplifier. Figure **11-6** suggests the basic principle.

In Figure 11-2, Oscillator 1 changes the frequency of its output when a conductive object moves toward or away from the yellow coil of wire, while Oscillator 2 changes the frequency of its output when you adjust variable capacitor C5.

The outputs from the oscillators go to the inputs of X3, the third XOR logic gate, which works as a mixer, as shown in Figure **11-7**. It compares the two oscillator signals and generates an audible signal if the frequencies differ.

Figure **11-8** shows how the mixed output depends on the frequencies of the two oscillators as they go in and out of phase with each other. When they are exactly or almost in phase, the mix has a generally low output. When they are exactly or almost out of phase, the mix has a generally high output. The variation between a low output and a high output happens so quickly, it creates its own audible frequency.

The idea of comparing two signals to generate a new signal at their difference in frequency is powerful and is often used in radios.

In radio, a circuit doing this is called a *mixer* or *frequency* mixer. (This should not be confused with an audio mixer, which simply adds several audio signals or perhaps allows fading between them. That's a different type of mixer.)

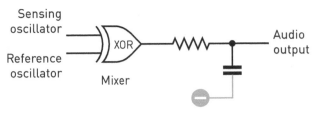

11-7 *The third XOR gate mixes the signals from the two oscillators.*

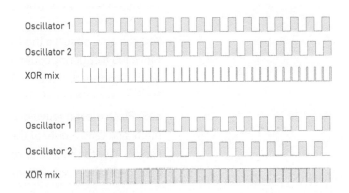

11-8 *How the third XOR gate mixes signals from the two oscillators.*

In the next experiment, you will construct a radio receiver where a mixer mixes the signal from the antenna with a signal from a tunable oscillator.

The circuit in this experiment will actually work with either NAND or XOR gates. Since the 4011B NAND chip and the 4030B/4070B XOR chip have the same pinouts, meaning that the supply-voltage pins and the gate inputs and outputs are in the same place, any of these chips will work.

MEASURING THE OSCILLATOR FREQUENCIES

If you still have the frequency counter that you built in Experiment 7, you can use it to measure the frequency of Oscillator 1 and Oscillator 2 in this experiment. Power the frequency counter through the USB port of your computer, as you did before. Connect the negative ground between the two breadboards, and connect the frequency counter input to Pin 11 of the chip, the output of the sensing-coil oscillator.

You should see about 2MHz. Take note of the value. Move the frequency counter input to Pin 3 of the logic chip, the output of the reference oscillator. You should see frequencies roughly in the range of 1.9MHz–2.1MHz, which will change as you turn the tuning wheel over its full range.

Notice that the whistle appears when the two oscillators are close in frequency and that when you change the frequency of one oscillator with the tuning capacitor, the whistle changes pitch.

You can also observe the frequency change when a metal object is detected. Move the frequency counter input back to Pin 11. Tune the metal detector again—connecting the counter will affect the oscillator. Make a note of the frequency reading with no metallic or conductive objects close to the sensing coil. Place a conductive and nonmagnetic object near the coil, listen to the pitch, and take a note of the frequency. Remove the metallic object, and replace it with the ferrite rod you used in Experiment 1. (You can leave the coils on.) Take a note of the frequency again.

You should see that the first conductive object increases the frequency, while the nonconductive but magnetic ferrite rod decreases the frequency. Either of these frequency shifts is audible in the earphones, but there is no way to tell them apart by listening. It sounds the same whether the sensing oscillator's frequency shifts up or down.

SENSING METAL AND MAGNETIC MATERIALS

If you measured the change in the oscillator frequency when the coil approached a conductive or a magnetic object, you saw that a conductive object increases the frequency, while a magnetic object decreases it. Both of these effects occur through changes in the sensing-coil inductance.

In Experiment 1, you learned that winding a coil on a magnetic material such as a ferrite rod increases the inductance of the coil. A larger inductance means the resonance frequency of the LC circuit is lower. So, with a magnetic material in the vicinity, the coil inductance increases and the frequency decreases.

In the AM receiver, the ferrite rod does two things: It increases the inductance of the coil wound on it, like here. It also concentrates the weak magnetic field—which is part of the radio wave you want to receive—so that the small coil becomes a much more effective antenna than the coil would be on its own. Both of these are due to the high magnetic permeability of ferrite, but only the inductance change is interesting here.

With conductive material, the situation is a bit more complicated. You can think of the conductive material and the sensing coil as forming a transformer. The sensing coil is the primary winding, where the oscillator creates an oscillating current. The conductive material acts as a short-circuited secondary coil, where the fluctuating magnetic field from the primary coil induces a current.

The current in the secondary coil creates a magnetic field of its own, in the opposite direction of the field from the sensing coil, as shown in Figure **11-9**.

The sensing coil senses the influence of the field from the conductive object. In total, the presence of the conductive object decreases the inductance of the coil, which increases the oscillator frequency.

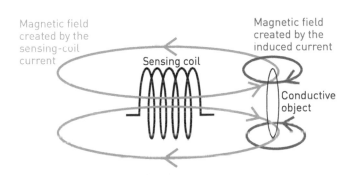

11-9 *The magnetic fields around the sensing coil and a nearby conductive object.*

TROUBLESHOOTING

If you cannot find a strong whistling sound when tuning the tuning capacitor, I can think of two things that might be wrong.

The first is that one of the two oscillators in the circuit is not working. If you have an oscilloscope or the frequency counter from Experiment 7, you can measure the signal or the signal frequency at Pins 11 and 3, the oscillator outputs. You should see a square wave at approximately 2MHz on both of them.

Second, the oscillator frequencies may be so far apart that the range of the tuning capacitor is not sufficient for tuning them to the same frequency. This may happen if your sensing coil has the wrong number of turns or is the wrong size. Again, a frequency counter is helpful for checking. You can also try adding a 220pF capacitor in parallel with either C2 or C3 to decrease the frequency of either oscillator.

EXTENDING THE SENSING RANGE

The range of sensitivity of the sensing coil is roughly as large as the diameter of the coil. This is related to the shape of the magnetic field lines around the coil. So, if you want to detect objects farther away, you need a larger sensing coil. However, the longer range of a large coil comes at a price: The magnetic field is spread over a larger area, which makes a larger coil less sensitive to smaller objects. So, you need to compromise with the coil size.

If you want to experiment, find an online coil-inductance calculator such as the one at

`66pacific.com/calculators/coil-inductance-calculator.aspx`

or use Wheeler's approximation from Figure 1-41. Experiment with choosing the number of turns and the coil diameter to keep the inductance roughly constant at 22μH. A larger coil needs fewer turns to have the same inductance. The oscillator frequency itself is not critical in this circuit, but the reference oscillator must be tunable to the same frequency as the sensing oscillator. A frequency counter is handy for experimenting. Once your larger sensing coil gives roughly the right frequency, you can tune the two oscillators by adjusting the capacitors—for example, by adding a 220pF capacitor in parallel with either C2 or C3, as suggested in the "Troubleshooting" section.

I suggest keeping the frequency below about 2.5MHz, since I found that the audio output from the mixer gets weak and unreliable at higher frequencies.

STATUS UPDATE

In this experiment, you saw (or heard) how the resonant frequency of a coil and capacitor circuit can be affected by its interaction with nearby objects, so long as they conduct electricity. Although this frequency can be far above the audible range, if you mix it with another frequency, the result can be easily heard.

In the next experiment, you'll discover how the concept of mixing can be used to build a wonderfully simple type of receiver that mixes a high frequency with radio signals.

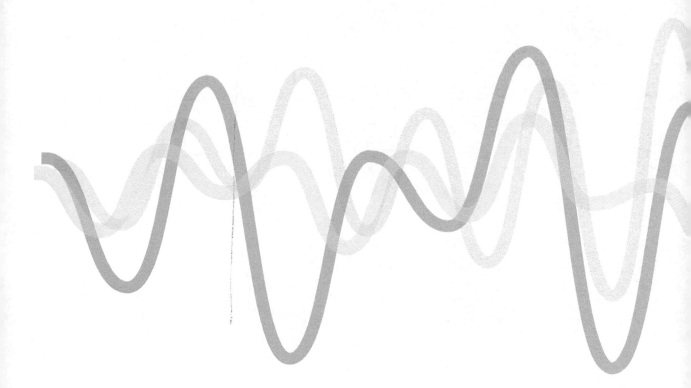

DIRECT-CONVERSION RECEIVER

In the previous experiment, you saw how a mixer circuit can create an audible frequency from two high frequencies when they interact. In this experiment, you will use capacitors and an inductor to build a tunable oscillator that creates its own high-frequency signal in the radio spectrum. When a radio signal of a similar frequency comes in through an antenna, the two frequencies mix to create an audible output.

This is called a *direct-conversion* receiver because it makes radio signals audible in a single step. It can receive Morse code transmission, or audio in a transmission mode known as *single side band*, or *SSB*, which is popular among radio amateurs on the shortwave bands.

You will also see how a filter works to select the frequencies that you want to deal with while blocking the rest. And you'll discover more about Morse code.

You Will Need:

- Resistors: 22 ohms (1), 100 ohms (2), 330 ohms (1), 1K (1), 2.2K (1), 47K (3).
- Trimmer, 10K (1).
- Inductors, 10µH (2), such as Bourns 78F100J-RC.
- Ceramic capacitors: 33pF (3), 68pF (1), 220pF (1), 470pF (1), 1nF (3), 47nF (1), 100nF (2).
- Electrolytic capacitors: 10µF (2), 470µF (2).
- Tuning capacitors, 200pF, type 223P (2)—the same as used in previous experiments.
- 2N3904 bipolar NPN transistor (1).
- Earphones and 1/8" audio jack, or speaker (1).
- LM386 amplifier chip (1).
- BAT48 Schottky diode (1).
- Slide switch, SPST or SPDT, to fit breadboard (1).
- 22-gauge wire for antenna and ground (at least 65 feet plus extra for ground). Copper tube can be used for ground connection instead of wire if desired. See the "Antenna" section below for further explanation on the required length.
- Optional: The frequency counter circuit from Experiment 7.

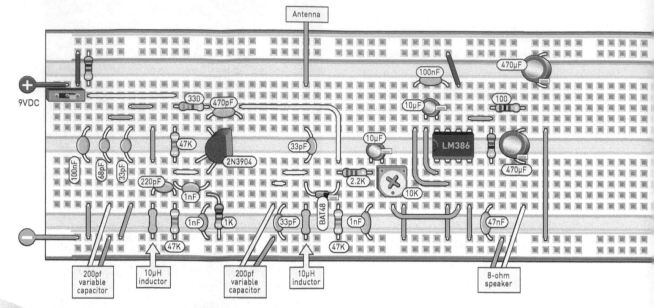

12-1 *The breadboard layout for the direct-conversion receiver.*

CONSTRUCTION

Construct the circuit shown in Figures **12-1** and **12-2**. The circuit uses a Schottky diode as a mixer and an LM386 audio amplifier, familiar from previous experiments. There are two variable capacitors: one for setting the oscillator frequency, the other for filtering the signals from the antenna before passing them to the mixer.

The grayed-out capacitor C14 in the audio amplifier is optional. When it's present, the amplification is increased from 20 times to 200 times. It's generally not necessary if you're using earphones, but for a speaker, it may be helpful.

ANTENNA

This receiver requires a long antenna and a ground connection, as in AMR1 in Experiment 1. The receiver is for the amateur-radio band, which you can find at 3.5MHz–3.8MHz, corresponding with a wavelength of 262 feet. One-quarter of the wavelength is a good length for a wire antenna. For this receiver, that means a 65-foot-long wire, ideally hung outdoors, high and free, with one end connected to the radio. For a ground connection, you can use a wire connected to a length of copper tube hammered into the

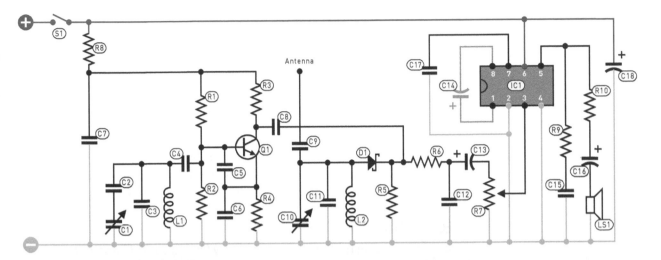

Resistors	
R1, R2, R5	47K
R3	330
R4	1K
R6	2.2K
R7	10K
R9	22
R8, R10	100

Capacitors	
C1, C10	200pF
C2	68pF
C3, C9, C11	33pF
C4	220pF
C5, C6, C12	1nF

Capacitors	
C7, C17	100nF
C8	470pF
C13, C14	10µF
C15	47nF
C16, C18	470µF

Additional Components	
IC1	LM386 amplifier chip
Q1	2N3904 bipolar transistor
S1	SPDT or SPST slide switch
L1, L2	10µH inductor
D1	BAT48 Schottky diode
LS1	8-ohm 3" or 4" speaker

12-2. Schematic of the direct-conversion receiver.

ground. If the ground is very dry, it can be difficult to achieve a good ground connection, in which case laying a piece of wire the same length as the antenna on the ground may be a good substitute. See also the antenna and grounding advice in Experiment 1.

The frequency is higher here than in Experiment 1, which makes the wavelength shorter. This means it's more practical to have an antenna a quarter-wavelength long. In cities, it may be difficult to set up an antenna, and the density of radio noise sources is higher. I tried using a ferrite bar antenna, as in AMR2, in this receiver but could not get it to work (perhaps because ferrite becomes more lossy and responds more weakly as the frequency increases).

LISTENING

Connect the antenna and ground wires to the radio. Turn the volume down. Apply power with the sliding switch.

Listen to the speaker or use earphones. Now, you have three controls to tune: the oscillator frequency, the filter frequency, and the volume. The general idea is that you tune the radio using the left-hand variable capacitor for the oscillator, then adjust the filter capacitor for the loudest signal.

Even more so than the AM broadcast band you listened to in previous experiments, the shortwave band is unpredictable and dependent on the time of day. Afternoon, evening, and night are the best times to receive stations; nighttime makes longer ranges possible. If you hear only static on your first attempt, just take a break and try again later in the day. Also, as with the AM broadcast band, local noise sources may hinder reception. See Experiment 2 for typical sources of noise and what to do about them.

RECEIVING MORSE CODE

For your first listening tests, I want you to try finding some Morse code transmissions, since they are easy to recognize. Start with the oscillator dial at the counterclockwise position. This is the lowest end of the frequency scale and is the part of the band where Morse code transmissions are concentrated (because of the **band plan**, an agreement about how the amateur-radio frequency bands are divided for different uses). Adjust the filter dial for the strongest signal. (There may be a peak in the clockwise-most setting, but try to find one in the middle-to-counterclockwise range of the dial.) Then, slowly tune the oscillator clockwise—upward in frequency. If you're lucky, you'll find one or more Morse code signals. As you tune past them, they will vary in pitch, much like the metal detector did when the reference oscillator was tuned across the frequency of the sensing oscillator. The receiver here is similar to the metal detector, but the search oscillator is now replaced by the signal from a distant radio transmitter picked up by the antenna.

CHECKING THE FREQUENCY

- If you have the frequency counter from Experiment 7, you can use it to check the tuning range. It should be roughly 3.5MHz–3.8MHz to match the amateur-radio frequency range.
- Connect the negative ground of the frequency counter to the negative ground of the receiver.
- Apply power to the frequency counter with a USB cable.
- Connect the frequency-counter input (remember the protective capacitor and resistor) to the oscillator output between Q1 and C8.

RECEIVING AUDIO TRANSMISSIONS

If you keep tuning upward in frequency, you get into the band allocated for voice transmissions. SSB audio is related to AM but is more efficient in the sense that it requires only half the **bandwidth**, the range of frequencies occupied by a single transmission.

The SSB transmissions are very sensitive to the tuning of the receiver. If the tuning is slightly off, the voice can be understood but the pitch is wrong. For larger tuning errors, the voice becomes garbled and unintelligible. It takes practice to get the tuning right, and on this receiver it can be quite difficult since a very small movement of the tuning wheel makes the difference between an intelligible signal and a garbled one.

There are several solutions offered by more sophisticated receivers: an additional fine-tuning dial, a multiturn tuning dial with some mechanical gears turning a variable capacitor very precisely, or a digitally controlled oscillator that enables you to select the tuning rate.

Another annoyance you will encounter with this receiver is that your hands affect the tuning by adding a tiny amount of capacitance. A solution here is to mount the tuning capacitors behind a metal plate connected to the negative ground of the radio circuit. You can also try to use the hand-capacitance effect to your advantage and fine-tune the reception frequency by moving your hand near the capacitor.

BROADCAST-STATION INTERFERENCE

As you tune over the band, you may also hear broadcast AM stations, accompanied by a whistling sound. The whistle comes from the carrier wave of the AM signal. There should be no AM broadcast stations in the amateur-radio frequency band, so probably the station is at a frequency that is a multiple of your oscillator frequency. In the mixer, these frequencies combine with the oscillator frequency and produce audio-frequency results. This is not wanted, and it's the task of the filter (C10, C11, and L2) to keep such frequencies out. However, the broadcast signals can be strong, and sometimes they are heard anyway. To see why the filter is needed, you can try removing C10, C11, and L2 temporarily, and you will likely hear a lot of signals from other frequency bands.

HOW IT WORKS

There are four sections in the circuit: an oscillator, a frequency filter, a mixer, and an audio amplifier. The main idea is that the mixer combines the signals from the antenna with a signal from the oscillator. As in the metal detector, this creates new frequencies. The frequency we want is the audible beat frequency between the radio station being received and the oscillator.

THE OSCILLATOR

The oscillator is constructed around the single transistor in the circuit. It uses an LC circuit consisting of C1, C2, C3, and L1 to set the oscillation frequency. Like the transistor oscillator in the AM transmitter in Experiment 3, this is a Colpitts oscillator. As I said before in Experiments 3 and 11, an oscillator can be seen as an amplifier with feedback from the output to the input that allows a selected frequency to go around and be amplified repeatedly. The capacitors C5 and C6 provide feedback from the amplifier output at the emitter to the amplifier input at the base, and the LC circuit selects the frequency.

Here, the capacitors and coil in the LC circuit are chosen to make the tuning range 3.5MHz–3.8MHz to match the shortwave amateur-radio frequency band we're trying to receive. An oscillator in a radio receiver or transmitter is often called a *local oscillator* because the signal it produces is used locally within the circuit.

THE FILTER

The filter consists of C10, C11, and L2, wired as a resonance circuit. It selects which frequencies from the antenna are passed through to the mixer. This filter is necessary because signals in other frequency bands—especially at integer multiples of the frequency we want to receive—can cause interference in the mixer.

THE MIXER

When dealing with audio, a mixer is a circuit that simply adds (or averages) signals—for example, with a voltage divider. In radio, a mixer can (and usually does) mean something different. It still combines two signals, but in a way that generates completely new frequencies. For example, a mixer can create an audible signal from two radio-frequency signals. That is how we used a mixer in the metal detector: to compare the signals from the two oscillators in an audible manner. There, the mixer was a logic XOR gate.

In the receiver in this experiment, the diode D1 acts as the mixer. I chose the same BAT48 Schottky diode here as in the crystal receiver in Experiment 1 because it has a low threshold voltage.

Here's an example to explain how a diode mixer works. Figure 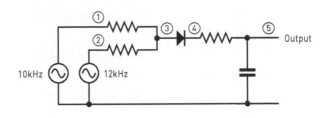 12-3 shows a diode connected to two signal sources, outputting sine waves at 10kHz and 12kHz. For simplicity, let's assume an ideal diode and ignore the threshold voltage. The input frequencies here are much lower than what is used in the radio, to make them easier to visualize.

12-3 A simple diode mixer connected to two oscillators generating sine waves of 10kHz and 12kHz. The signals pass through two equal resistors and then through a diode. The resistor and the capacitor to the right of the diode form a low-pass filter.

Figure 12-4 shows the signals at several points in the circuit. The two resistors create an average of the two signals at point 3. When the two input signals are in phase so that the peaks and troughs line up, the signal at point 3 has a large amplitude. And when the input signals are out of phase so that peaks in one line up with troughs in the other, the amplitude is small. Since the two signals have different frequencies, they will drift in and out of phase with each other, so that their average at point 3 varies in amplitude. When listening to such a signal, the ear interprets it as a single tone varying in strength, particularly when the two frequencies are within a few hertz of each other so that the amplitude varies slowly. The phenomenon is called a **beat**, which has a **beat frequency**. The closer in frequency the two input signals are, the slower they drift in and out of phase with each other. The beat frequency is the **difference** between the frequencies of the two input signals.

Next, the average signal is passed through a diode, which cuts off the negative parts of the signal. As in Experiment 1, where a diode was used to demodulate an AM signal, the signal can

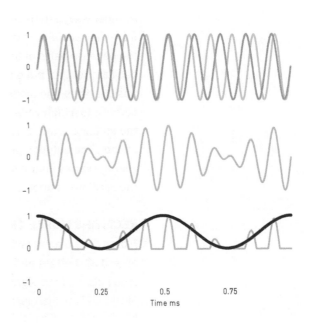

12-4 Waveforms at different points in the circuit in Figure 12-3. The top panel shows the sine waves from the two oscillators at points 1 and 2. At point 3 (middle panel), the voltage is the average of the two input signals. In the bottom panel, we see that at point 4, after the diode, the negative periods of the signal have been cut off (shown in blue). Smoothing the output from the diode gives a slow sine wave signal at point 5 (shown in black).

now be low-pass filtered, which creates a new smooth signal at a lower frequency than either of the input signals, shown in black in Figure 12-4.

(Notice how the XOR gate output in Figure 11-8 functions similarly to the diode, detecting how the two signals go in and out of phase with each other and generating a low-frequency signal.)

Mixers can also produce a signal at a higher frequency than the input signals—namely, at the sum of the input-signal frequencies. Typically, a mixer outputs a combination of the sum frequency, the difference frequency, and sometimes the two original signals. You can then use a filter to select only the signal you want among these mixer products. In the receiver (see Figure 12-2), R6 and C12 form a low-pass filter selecting only audio frequencies.

Another way to think about the diode is as a switch. In the receiver, the signal from the oscillator is much stronger than the one from the antenna. During the negative parts of the oscillator signal, the diode conducts (because the oscillator output pulls the diode cathode to a low voltage), and during the positive parts, the diode does not conduct (because the oscillator pulls the cathode to a high voltage). When the diode conducts, small fluctuations from the antenna signal can get through, but when the diode is blocking, they cannot. The output signal from the diode is averaged by R6 and C12. The average varies as the signal from the antenna drifts in and out of phase with the oscillator signal, similar to Figure 12-4.

RECEIVING MORSE TRANSMISSIONS

Morse code transmissions consist of a single sine-shaped signal switched on and off. Letters and numbers are encoded in the rhythm of short and long pulses of the signal. An AM receiver is not well suited to receive Morse code transmissions, because both when the signal is on and when it is off the amplitude is constant—and then the AM receiver detects a constant amplitude and stays silent. The direct-conversion receiver in this experiment does a much better job with Morse. The trick is to tune the local oscillator to a frequency near (within perhaps 2kHz) but not exactly at the same frequency as the Morse transmission, as shown in Figure **12-5**. Then, whenever the signal is on, the receiver will produce a beep. This is the typical way of receiving Morse transmissions: a series of long and short beeps.

If you can hear multiple stations transmitting at nearby frequencies, things get more confusing. The receiver in this experiment picks up signals both

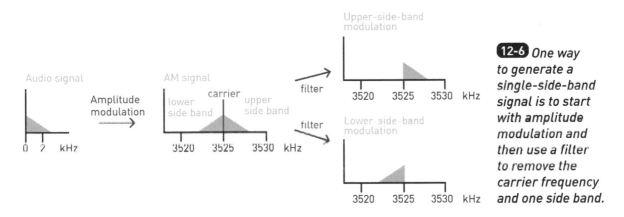

12-5 *Diagram of receiving signal A at the frequency 3,525kHz with a direct-conversion receiver. With the local oscillator LO at either 3,527kHz or 3,523kHz, the result is a 2kHz tone. The right-hand image shows that when two signals A and B are present, both appear in the output audio, with frequencies determined by their distance to the local oscillator frequency.*

12-6 *One way to generate a single-side-band signal is to start with amplitude modulation and then use a filter to remove the carrier frequency and one side band.*

above and below the local oscillator frequency, as shown on the right side of Figure 12-5. In more sophisticated receivers, it is possible to select only frequencies above or below the tuning frequency.

AUDIO TRANSMISSIONS WITH SSB MODULATION

To explain SSB, we have to look at the spectrum of frequencies present in an AM signal.

If you have an AM transmitter and turn it on but do not speak into the microphone, it will transmit the carrier wave with a constant amplitude. In a spectrum, this signal has a single frequency. If you now add an audio signal (frequencies of 0kHz–3kHz are typical for speech over radio), the carrier amplitude is varied in time. In the spectrum, this shows up as additional signals on both sides of the carrier frequency, called **side bands**. The spectrum of an audio signal is shown in Figure **12-6**. (In this example, it's shaped like a triangle.) The amplitude-modulated version of this signal has one such side band on either side of the carrier frequency. The two side bands contain the same information. The idea in single-side-band modulation is to transmit only the frequencies in one of the side bands. One way to do this is to filter out the carrier frequency and one of the side bands from an AM signal. "Upper-side-band modulation" and "lower-side-band modulation," abbreviated as *USB* and *LSB*, refer to which side band is kept.

One drawback with SSB is that it cannot be received with the simple receivers you built in earlier experiments. It requires a receiver with a local oscillator, which in a way replaces the filtered-out carrier frequency. Figure **12-7** shows the direct-conversion receiver tuned to an LSB signal. The correct tuning is to have the local oscillator just above the LSB signal since this gives back the original audio. Slight shifts in the tuning give comprehensible audio with a pitch shift. Tuning the local oscillator below the signal results in audio in which the high and low frequencies have changed place, which sounds strange (and has historically been used as a form of analog speech encryption).

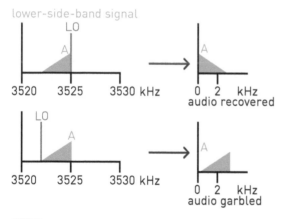

12-7 *Diagram of receiving a lower-side-band audio signal and the resulting audio output for two different local-oscillator frequencies. When the local-oscillator frequency is just above the single-side-band signal, the resulting audio matches the original audio signal. With the local oscillator just below the signal, the signal can be heard, but in a garbled way, because low and high frequencies have changed places.*

PERFORMANCE

How good is the receiver? I was surprised it works this well, constructed on a breadboard, which isn't really recommended for radio-frequency circuits. It has some weak points: The LM386 amplifier is a bit noisy, and you notice this when listening to it with headphones. (Try disconnecting the mixer diode from the amplifier—the noise is still there!) The local-oscillator tuning changes when your hands are close to it. Mounting the tuning capacitor and the oscillator part of the circuit inside a separate metal enclosure would help here. The original VK6 80 design I based this receiver on (see sources below) recommended a metal enclosure and is probably better in other respects, too. I made different choices in order to keep the build as simple as possible and to use the same components as in the other experiments. I think there is a lot to learn from building simple radios, even if the performance isn't the best. Plus, they allow you to better appreciate the design choices in more advanced radios.

Hopefully, you could hear some radio amateurs on the shortwave band. In the next and final chapter, I will suggest some ways to continue with radio as a hobby. One of them is to get an amateur-radio license so that you also can transmit on the shortwave amateur-radio bands.

SOURCES AND FURTHER READING

The oscillator and mixer parts of the circuits are based on the VK6 80 receiver by Peter Parker, which was featured in the September 1995 issue of *Amateur Radio*, published by the Wireless Institute of Australia. You can read the article on page 486 at:

worldradiohistory.com/AUSTRALIA/Amateur-Radio/Amateur-Radio-AU-1995.pdf

"Direct Conversion Receivers—Some Amateur-radio History", by Wes Hayward, who in 1968 popularized the design still used in many amateur radios today:

w7zoi.net/oldtech/dcrx68a.pdf

Here are a couple of other good articles on the subject: "High-Performance Direct-Conversion Receivers," by Rick Campbell, published in the August 1992 issue of *QST*. This is quite advanced, with interesting ideas on what makes a direct conversion receiver good. You can read the article online at:

arrl.org/files/file/Technology/tis/info/pdf/9208019.pdf

STATUS UPDATE

In the preceding 12 experiments, I've introduced you to the aspects of radio that are easiest to understand, and I've shown how accessible the radio spectrum is if you assemble just a few affordable components. You've seen that radio signals can be created very easily, and you've probably discovered that there are AM stations around you that you were unaware of.

The next and last section of the book will show you how you can get involved in the world of licensed amateur-radio transmissions, as well as some other ways to continue exploring the world of radio.

13

GOING FURTHER WITH RADIO

ndividuals can interact with each other instantaneously, all over the world, via the internet. In this era of instant global communication, you might assume that amateur-radio has become irrelevant, but this is not the case. The field endures; in fact, its signals are all around you.

The little transmitter circuits described earlier in this book are severely limited in power, to comply with government regulations. If you study to obtain a license, you can operate equipment with much greater range. It will literally open up a world of new possibilities.

This final chapter will describe call signs, communication protocol, QSL cards, frequency bands, and sources you may go to if you want to buy a low-cost transmitter and receiver. You'll also learn about software-defined radio, which takes advantage of the power of a desktop computer. In many ways, the world of amateur-radio has never been more easily accessible. I will also give you some suggestions for how to continue with building radio circuits, and show some useful test equipment.

You Will Need:
- Optional: A computer and an SDR receiver (1). See "Software-Defined Radio" on page 209 for details.

AMATEUR RADIO

Amateur radio, also called **ham radio**, is a way for amateurs to communicate with each other over radio frequencies. Several frequency ranges are allocated for amateur-radio use. To make your own transmissions on these frequencies, you will need an amateur-radio license. The details vary by country, but generally, getting a license requires passing tests to demonstrate your knowledge of rules, regulations, and technical requirements for operating radio transmitters properly.

Some aspects of the amateur-radio hobby include the following:
- Making contact with people who share your interests.
- Developing technical skills.
- Building and experimenting with circuits.
- Communicating worldwide, perhaps on equipment you build yourself.
- Emergency preparedness—communicating with other people when other methods fail, such as during a natural disaster.

Most countries have amateur-radio associations, which can give you more information about obtaining licenses. In the United States, the largest association is ARRL, the **American Radio Relay League**:

`arrl.org`

Amateur-radio associations publish magazines such as *QST* and *CQ*. These can be good sources of technical articles and construction projects.

LICENSING

In the United States, the FCC (Federal Communications Commission) proctors multiple-choice exams and grants amateur-radio licenses. There are three license classes: Technician, General, and Extra. Each class requires more knowledge while allowing you more rights, such as being able to use higher power or access more frequency ranges. For more information on the license classes, go to

`arrl.org/getting-licensed`

The topics covered in the license exam include the following:
- Rules that limit where and how you are allowed to transmit.
- Abbreviations used in radio communication.

- Basic electronics theory, such as units, components, and schematic symbols.
- Theory of radio waves and how they propagate.
- Different modulations.
- Antennas.
- Electrical safety.

If you are interested in getting a license, I suggest finding a local amateur-radio club where you can learn more. Try the club search at

`arrl.org/find-a-club`

You can find practice tests available on various websites, such as

`hamexam.org`

The ARRL website also has a question pool and a practice test. It requires registration but is otherwise free.

Passing a Morse code test used to be required in many countries, especially for access to the shortwave bands, but this requirement has been dropped. However, Morse code continues to be popular, since it can enable communication under conditions when voice transmissions are not possible.

Besides Morse code and voice communication, there are more modern digital modes, which send text over radio. One way to use these is to connect a computer sound card to a radio to send and receive the digital signals. You can also send and receive Morse code this way so that you can encode and decode messages without learning to send and interpret Morse manually.

CALL SIGNS

Every amateur-radio station is identified by its own unique **call sign**, consisting of letters and numbers. An example is my call sign, OH1HSN, in which *OH* is the country code for Finland, the number 1 identifies a region in the country, and the letters *HSN* identify a particular station. In the United States, call signs begin with the letters *A*, *K*, *N*, or *W*.

In the United States, a transmitting station's call sign should be sent at the beginning of each transmission, and then once every 10 minutes. The required repeat interval varies by country.

When you speak the characters in a call sign, you use the phonetic alphabet. *A* is spoken as "Alpha," *B* is "Bravo," *C* is "Charlie," *D* is "Delta," and so on. These words are chosen to sound distinct so that they can be understood even though the connection may be poor. Some radio operators may use different phonetic alphabets, especially for local communication in a language other than English.

COMMUNICATION PROTOCOL

Two-way communication typically begins with one station sending a general call such as "CQ CQ CQ, this is OH1HSN."

You can think of CQ as "seek-you." The call sign is then spelled out using the phonetic alphabet, which in this case would be "Oscar Hotel One Hotel Sierra November."

Someone receiving this may reply, "OH1HSN, this is OH1AA," again with the call signs spelled out. If the first station calling hears this, they will usually send a longer message. It is customary to exchange signal reports, which consist of two or three numbers rating the other station's signal strength and how well it can be understood, followed by the station location. Locations are generally given with city names or with coordinates in a brief format consisting of two letters, two numbers, and two more letters.

Beyond that, you can talk about anything on amateur radio. It's a matter of personal preference whether you engage in long conversations or you are just seeing what you can find.

QSL CARDS

People in amateur-radio often enjoy the challenge of communicating over very long distances, or with as many different stations or countries as possible. As proof of a radio contact, radio amateurs exchange **QSL cards**.

The cards, which look like postcards, mention the call signs of the two stations, the date and time of the contact, the frequency used, and the modulation (such as Morse, audio, or digital). Some examples of call cards are shown in Figure **13-1**.

The ARRL also provides an electronic version of call cards, known as the Logbook of the World.

FREQUENCY BANDS

Several frequency ranges, or bands, are allocated for amateur use. The exact ranges, power limits, and regulations vary by country, but fortunately, the bands overlap reasonably well, making international communication possible.

Broadly speaking, the bands are divided into shortwave (3MHz to 30MHz) and higher frequencies (VHF and UHF).

Radio signals in the shortwave bands can be reflected back to Earth by certain layers of the atmosphere. This makes communication far beyond the horizon possible. However, solar radiation affects the atmosphere's properties throughout the day, so precisely which communication paths are open varies. This is called **radio wave propagation**. The different bands in the shortwave range can also have very different characteristics at different times.

13-1 *QSL cards from various amateur-radio stations.*

Some of the popular bands are listed in the following table. The wavelength is approximate, but it is often used to refer to the band. There are more amateur bands both lower and higher in frequency.

Wavelength	Frequency	Band
80m	3.5MHz–4.0MHz	shortwave
40m	7.0MHz–7.3MHz	shortwave
20m	14.0MHz–14.35MHz	shortwave
10m	28.0MHz–29.7MHz	shortwave
2m	144MHz–148MHz	VHF
70cm	420MHz–450MHz	UHF

For local communications, the 2m and 70cm bands are suitable. They are reliable for local communication, with a range of tens of miles for handheld radios and up to a hundred miles with larger radios and with antennas mounted high up. The range is limited, since the radio waves are generally not reflected back to Earth by the atmosphere. The limited range has an advantage in that distant stations will not interfere with your local communications.

For amateur communication on the VHF and UHF frequency bands, there may be local *repeater stations*, which boost the range of your transmissions. The repeater works by listening on one frequency and retransmitting on another frequency. Such repeaters are a way for the local amateur-radio community to stay connected.

The VHF and UHF bands are generally divided in *channels*, with specific frequencies you are encouraged to use, varying according to your country and region. The shortwave bands generally don't have such channel divisions.

I like the shortwave bands because exploring them can feel like a treasure hunt. They are unpredictable, but very long connections are possible. You can use home-built equipment or buy off-the-shelf hardware.

OBTAINING A RADIO

A radio that can both transmit and receive is called a *transceiver*. Another option is just to buy a receiver, which you can use without any need for a license. Software-defined radio is another possibility.

Before you invest in any equipment, think about what kind of communication you are interested in, and perhaps try to visit a local radio club, where you can get advice. A club may also enable you to try out some radios and perhaps buy secondhand gear.

Some amateurs find it interesting to use vintage communication equipment, such as surplus military radios, which may contain vacuum tubes.

SOFTWARE-DEFINED RADIO

Software-defined radio, or SDR, is equipment where a part of the signal processing is done digitally using a computer. This means that new modulations can be added by installing software applications, unlike traditional radios where, all functions are implemented in hardware.

A *scanner* is a piece of radio equipment that can tune over a wide frequency range, searching for interesting signals. With SDR, you can do this on a computer screen, which gives you the advantage of seeing nearby active frequencies instead of just searching by listening.

The RTL-SDR is a small SDR that was first used as a cheap receiver for digital TV broadcasts, connecting with a computer that handles a part of the demodulation of the signal and displays the video. Some people who wanted to use this receiver with Linux wrote their own driver program for it. They discovered that besides receiving digital TV, the receiver can be tuned to any frequency between about 20MHz and 1,700 MHz, converting a section of a few megahertz into a digital form that can be sent over a USB port. Then a computer program can process the data further to receive many different kinds of transmissions.

If you live near an airport, you can pick up the communication between the air traffic control and the aircraft. This usually takes place in the 118MHz–137MHz frequency range using AM modulation. Near harbors and shipping routes, you can listen to the marine VHF band, which covers 156MHz–174MHz and uses FM.

There are also interesting digital transmissions you can receive. Ships and airplanes carry automatic transmitters that periodically transmit their position, speed, and heading and an identifying number. By receiving these signals, you can "see" nearby vessels in real time. For ships, the system is called the *automatic identification system (AIS)* and operates at frequencies 161.975MHz and 162.025MHz in the marine VHF band. For aircraft, the system is called *ADS-B* and operates at 1,090MHz.

On websites such as Marinetraffic.com (ships) and Flightradar24.com (aircraft) you can see this information gathered by a network of stations and plotted on a map. Still, I think it's more interesting to receive the signals yourself.

Before using scanner radios or SDR receivers such as the RTL-SDR, do check your local regulations, since they vary by country, and there may be legal limitations on what you can receive—especially if you consider publishing information you have received over radio.

HARDWARE

There are many different SDR devices for different purposes. Some can only receive, while the more expensive ones can also transmit, signals. One inexpensive but still capable model I can recommend is sold by RTL-SDR.com. I recommend buying through this website (I have no relationship with it), as there are multiple clones, some of which are actively misleading. What sets the more official receiver apart from the generic clones is better frequency accuracy and the possibility to also receive the lower frequencies of the shortwave band. The website's SDR receiver includes a telescoping dipole antenna.

SOFTWARE

I will mention three free programs. Their installation may be more difficult than that for average programs, and using the programs does entail a learning curve. You can expect to spend some time reading the manuals.

GQRX is available from Gqrx.dk and runs under Linux or macOS. This is a general receiver for different modulations: FM, AM, SSB, and Morse.

SDRangel can be found at SDRangel.org, running under Linux, macOS, and Windows. This is especially recommended for digital modes such as the ship and aircraft tracking mentioned above. When you receive either of those transmissions, the program shows a map window where the location of the source is displayed.

SDR# (pronounced "SDRsharp") is available from Airspy.com and runs under Windows only. This is a general-purpose receiving program.

ANTENNA

For good results with SDR or any radio, you do need an antenna. With the RTL-SDR radio, you can buy a telescoping dipole antenna. Cheaper RTL USB devices often come with a small antenna on a magnetic base. This can probably pick up the local FM broadcast stations but not much else. Whatever antenna you have, if you can mount it outdoors, high, and free from interference, you will have a better reception than indoors.

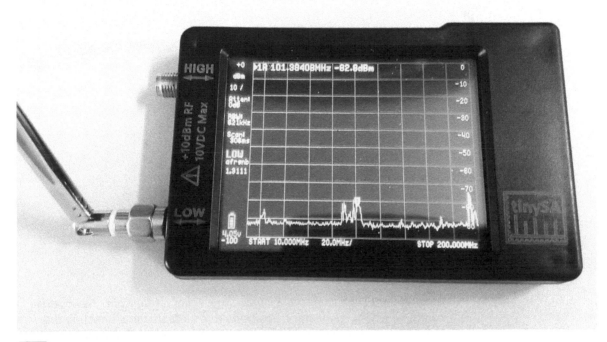

13-2 *The tinySA spectrum analyzer. The yellow curve shows the signal strength at different frequencies, currently set to show 10MHz–200MHz. The peaks in the middle of the screen show stations in the FM broadcast band.*

TEST EQUIPMENT

In the experiments, you already built two pieces of test equipment: a signal generator and a frequency counter. If you enjoy building radio circuits, consider acquiring two additional instruments to be able to test and measure radio signals: an *oscilloscope* and a **spectrum analyzer**. Oscilloscopes are described in Appendix D, as they have multiple applications in electronics. A spectrum analyzer, at first glance, is similar to an oscilloscope. It measures a signal and displays it on a screen. But where the oscilloscope displays the signal in time, the spectrum analyzer displays it in frequency.

Spectrum analyzers used to be laboratory equipment (with a price tag to match). There is a new one called tinySA shown, in Figure **13-2**, with a much more hobbyist-friendly price. You can find it at TinySA.org. The tinySA can also generate test signals, which is useful when testing receivers.

13-3 *Some useful cables and connectors. From left to right: Two coaxial cables with SMA connectors; an attenuator, suitable for connecting the output of a low-power transmitter directly to the spectrum analyzer; and a homemade SMA-to-breadboard adapter, made by soldering a two-pin header to an SMA connector.*

As with the RTL-SDR above, I advise only buying from stores mentioned at the official website, to avoid clones of unknown quality.

You should also consider getting a **30dB attenuator** (which should only cost you a few dollars) if you want to connect any transmitter directly to the analyzer. The spectrum analyzer cannot handle high input powers, and it may be overloaded or show spurious peaks if the signal is too strong. The attenuator is connected between the transmitter and the spectrum analyzer to reduce the signal to a safe level. Note that *dB* is an abbreviation for **decibels**, a measurement of how much the signal is attenuated. *30dB* means $^1\!/_{1,000}$ of the input power is transmitted to the output.

The tinySA uses connectors called **SMA**. These are common on **coaxial cables** for various instruments and radios when the signal powers are low and the cables can be thin. RTL-SDR radios and many other similar receivers also use SMA connectors. Figure **13-3** shows some cables and connectors.

If you build a transmitter, you can use a spectrum analyzer to look at the signal it produces. In particular, it will show if you are generating signals, such as harmonics, at unintended frequencies.

You can connect an antenna (a telescoping rod or a small wire loop) to the input. Walk around and search for sources of radio noise in your home. I found that a drawing tablet for a computer emits a strong signal at about 660kHz. I also heard this when testing the AM receivers.

Using an antenna, you can also look for interesting signals to listen to. The frequency range you can watch at once is much broader (hundreds of megahertz) than what an RTL-SDR can show at one time (a few megahertz).

PHYSICS AND MATHEMATICS

Oscillators, resonances, and spectra are phenomena that turn up in many physical objects, such as pendulums, the vibrating strings in musical instruments, and organ pipes. In physics, a simple model for vibrating systems is the **harmonic oscillator**. This model is helpful for understanding LC circuits in radios. Waves are important in many phenomena, including sound, waves on a water surface, or the wave function describing particles in quantum mechanics. In this book, the most important waves are in the electromagnetic spectrum, including visible light, X-rays, and, of course, radio waves.

A point where mathematics is useful for radio is in describing, analyzing, and processing signals. Useful concepts include **trigonometry** (for the sine function) and **logarithms** (for the decibel unit). More advanced topics include **complex numbers** and **Fourier analysis** (for describing how a signal can be decomposed in sine waves of different frequencies).

Mathematical signal processing on a computer, called **digital signal processing**, is what enables software-defined radio.

The more technical knowledge you acquire, the better you will be equipped to understand and explore the fascinating topic of radio.

MICROCONTROLLER WIRELESS DATA COMMUNICATION

In Experiment 9, you learned about remote controls on the 433MHz frequency. If you want to, for example, set up a network of wireless sensors and want to transfer more than a few bytes of data at a time, or you require the data to be transmitted reliably, the remote control modules become impractical. Then you can explore the following options:

WiFi: Many microcontrollers support WiFi for wireless networking. The Raspberry Pi Pico, recommended in this book, doesn't, but the closely related **Raspberry Pi Pico W** does. WiFi is fast but quite demanding of power, and the range is limited (perhaps 300 feet under ideal conditions). You don't need to know very much about radio to use WiFi on a microcontroller. It is more a matter of computer networking and of programming the network chips on the microcontroller module.

Other radio modules: There are several radio modules that can be used with microcontrollers—for example, based on the CC1101 chip. These are much more sophisticated than the remote control modules but also more involved to use and to program. Such modules typically use license-free frequency ranges in the same way as the remote control modules in Experiment 9. For the software side, a good starting point is the Arduino library RadioLib, at

github.com/jgromes/RadioLib.

In the library documentation, you can find a list of compatible radio modules.

LoRa, which stands for "long range," is a specific (and proprietary) communication standard for which dedicated modules exist, promising long range (miles) with low power consumption when the amount of data to be transmitted is limited to a few kilobytes per second or less.

BUILD MORE RADIO CIRCUITS

Filters, oscillators, amplifiers, and mixers are the central concepts of radio construction. Almost any radio circuit will contain them as building blocks, and you have seen them all in action in the projects in this book. Here is a small summary and some terminology for further studies:

Filters allow signals of some frequencies to pass while, blocking others. You have seen *low-pass filters* in the audio circuits in the book. LC resonance circuits can act as *band-pass filters*, allowing only signals close to the resonance frequency. The resonance circuits in AMR1 and AMR2, and the filter in the direct conversion receiver in Experiment 12, are examples of this. In radios, filters can be built using resistors, capacitors, and inductors, but there are also specific filter components—for example, *ceramic resonators* and *SAW filters* (*SAW* stands for "surface acoustic wave"). Especially for audio and other low-frequency signals, good filters can be constructed with *operational amplifiers*.

An *amplifier* boosts the amplitude of a signal. They are used in both receivers and transmitters and are usually constructed with transistors or integrated circuits. Receivers may contain an amplifier stage to boost the weak signal from the antenna before further processing and another amplifier, which makes the final audio signal strong enough to drive a speaker. Experiment 2 featured both kinds of amplifiers. In transmitters, an amplifier may boost the signal to the antenna, as the transistor did in the final version of AMT2B in Experiment 6.

An *oscillator* creates a signal at a specific frequency. Experiments 1 and 2 used oscillators built around the 7555 timer chip. Experiments 3 and 12 contained an oscillator constructed from a transistor and a resonance circuit. In general, many oscillators can be thought of as an amplifier with positive feedback through a filter.

In Experiment 5, you used the Pico as a frequency generator, dividing the clock frequency to create a lower, precisely known frequency. A limitation of this approach is that not all frequencies can be generated, only integer fractions of the Pico clock frequency. There are clock-generator circuits that use a technique called a *phase-locked loop* to generate a precise and controllable frequency. A popular one is Si5351A, which is a chip so tiny that it is difficult to handle on its own, but it is available on breakout boards from Adafruit. It is controllable with I2C, so it should work well with the Pico (although programming it appropriately will require some research). How about using such a frequency generator as the local oscillator in the Direct-Conversion receiver of Experiment 12? Then you could have a precise and stable frequency controlled by the Pico, with frequency readout on an LCD screen. Perhaps a *rotary encoder* would be nice as the tuning control.

Mixers in radios allow moving a signal from one frequency to another. You saw a diode acting as a mixer in Experiment 12 and an XOR gate in this role in Experiment 11. There are more types with different properties—for example, *diode ring mixers*, available in packages resembling IC packages.

Another approach that has become popular in hobbyist designs of the last decades is to use an analog multiplexer chip as a mixer, often in a circuit called a *quadrature sampling detector*. This mixer is a good fit for a software-defined radio, where the analog signal is digitized after the mixer and the final demodulation and processing happens in software.

Being familiar with these radio building blocks, you can look for more radio circuits to construct. Try the books and web pages I've mentioned in the previous chapters or in some of the amateur-radio magazines mentioned above. You can also try assembling a kit. I recommend the SoftRock series

of software-defined radios (I once built one, it worked, and the instructions were thorough), available at www.softrock.com, and I have seen several promising kits from QRP Labs (which I have not tried). Many of these use the quadrature sampling detector design. These are more advanced than the projects in this book and require soldering. Several of the projects are transceivers for the amateur bands and require an amateur license to operate, but there are receive-only options too.

I want to give one final piece of advice: Do not make it too complicated from the start by trying to get all possible features in the radio you are planning to build. Start simple and build something. Good luck, and have fun!

APPENDICES

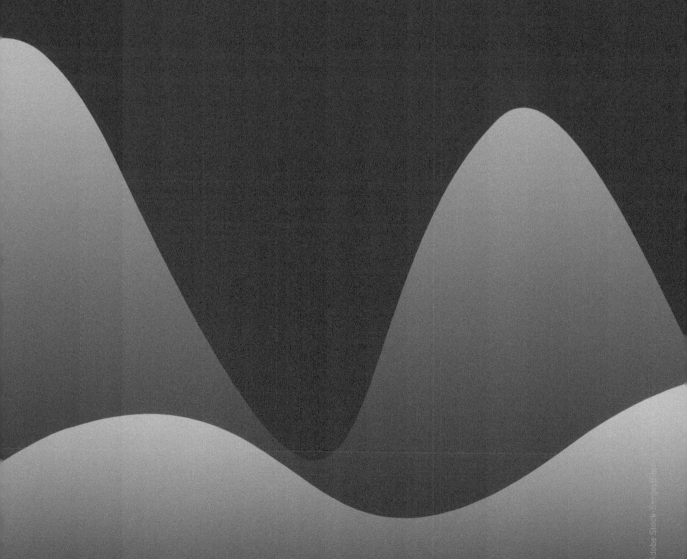

APPENDIX A: COMPONENT SPECIFICATIONS

This appendix contains information about the components used in the experiments in the book.

At the time of writing, kits are in preparation, but I do not know yet exactly which components they will contain. Please visit these two sources and search for "Make:Radio":

- www.protechtrader.com
- www.makershed.com

You can also order the components yourself, but you will likely have to order from several places. The sources are listed in Appendix B.

The table lists the components needed for each experiment. The column headed "Minimum" shows the minimum number of each component you need to perform the experiments, assuming you disassemble one before building the next, so that you can reuse the components. The column headed "Total" shows a recommended set of components, which allows you to have several circuits assembled at once. Specifically, I think the following circuits are fun together, and these are possible with the components in the total column:

- Experiments 2 and 3: AMR2 receiver and AMT1 transmitter
- Experiments 2 and 6: AMR2 receiver and AMT2B Pico transmitter
- Experiments 3 and 7: AMT1 transmitter and frequency counter
- Experiments 7 and 11: frequency counter and metal detector
- Experiments 3 and 10: AMT1 transmitter and regenerative receiver
- Experiments 6 and 10: AMT2B Pico transmitter and regenerative receiver

In Experiment 9, to simultaneously build the remote control receiver and transmitter requires two Picos, two breadboards, and two USB cables.

RESISTORS

The resistors used in this book are the common ¼ watt type. I strongly recommend using resistors with 5% tolerance because their four-band color code is easy to read: two bands for leading digit, one band for the number of zeros, and the fourth, gold-colored band for tolerance. (See Appendix C for interpreting the color code.) The more accurate resistors have a five-band code that is much harder to read.

Resistors are so cheap, you may find it worthwhile to buy a hundred of each of the values listed in the table. You can also consider buying a resistor set, which typically contains 10 or 20 resistors each for a set of common values. Such a set will likely contain all the values needed for this book, but check before you buy, and make sure the resistors have 5% tolerance. Resistor sets are available from eBay, Amazon, and many other sources. Good search terms are "carbon film resistor" and "through-hole" (to avoid surface-mounted resistor variants, which will not fit on a breadboard). One suitable resistor type is Yageo CFR-25JR-52-*330R*. (The italicized segment of the product number indicates the resistance.)

TRIMMERS

Ideally, the trimmers (more properly known as trimmer potentiometers) should have straight pins to fit nicely in the breadboard. (The Vishay T7 series, the Bourns 3306F, K, P, and W series, and the Bourns 3362F, H, P, and R series are examples.) If you use trimmers with kinked pins, you may need to squeeze them straight with pliers. (This topic is explored at greater length in the book *Make:Electronics*.)

Trimmers are used as volume controls in several experiments. Adjusting them requires a screwdriver, so if you are building a permanent version of a circuit, you may wish to substitute a panel-mount **potentiometer**, which is electrically equivalent to a trimmer but physically larger, with a shaft that can be fitted with a knob. If you buy a full-size potentiometer, you should not expect it to fit on a breadboard.

For volume-control purposes, a **logarithmic potentiometer** might be preferable, as it produces even variation of loudness over the rotation range. For breadboard experiments, however, the common linear trimmer potentiometers work well enough.

CERAMIC CAPACITORS

For buying ceramic capacitors, the two main things to consider are clear markings and the **dielectric material**, which acts as the insulation. Ceramic capacitors generally have their value marked with a three-digit code. (See Appendix C.) However, on some capacitors, the markings are so tiny, they cannot be read without magnification. (Use a magnifying glass, a loupe, or a smartphone camera.)

Ceramic capacitors contain tiny, thin metal plates separated by a ceramic dielectric material. There are several types of dielectric materials with different properties. The choice is a trade-off between price, physical size of the capacitor, and stability when the conditions change.

The cheapest materials may have a high temperature coefficient, meaning that the capacitance varies a lot as the temperature changes. For some dielectric materials, the capacitance may further change when a voltage is applied! When the capacitor is used to smooth a power supply, the exact capacitance may not matter much as long as it is large enough. But in radio circuits, especially in LC circuits, the component values may be chosen to give an exact resonance frequency. For example, in Experiment 3, the transmission frequency is set by a ceramic capacitor and a coil, and in Experiment 12, the reception frequency is set by a circuit made of a coil and several capacitors in the local oscillator.

I believe that for the breadboard experiments in this book, you can get the circuits working even with the cheapest capacitors. But if you have a choice, try to avoid the most unstable dielectrics. For that, you need to know some of the dielectric types and their codes. The most common ones follow:
- C0G or NP0: very stable; available for values up to about 1nF.
- X5R or X7R: moderately stable; good choice for capacitances above 1nF.
- Y5V: the cheapest option and the material enabling the smallest physical size, but capacitance varies strongly with temperature (and often with the applied voltage).

ELECTROLYTIC CAPACITORS

The circuits in the book require three kinds of electrolytic capacitors. They should be "through-hole" and "radial" (as opposed to "axial" where the two leads are attached to opposite sides of the capacitor body). Electrolytic capacitors have a maximum voltage rating. 16V is a sufficient maximum for the circuits in the book. Higher voltage models will work, they just tend to be larger and more expensive.

INDUCTORS

Inductors with three different inductance values are specified in the parts list. The recommended type looks like a resistor and has a color code in the same style, giving the inductance in microhenries. For example, the following inductors from Bourns are suitable: 78F100J-RC (10µH), 78F220J-RC (22µH), 78F102J-RC (1000µH).

FERRITE ROD

The ferrite rod used in the experiments is ⅖" (1cm) in diameter and 6-⅓" (16cm) long. Longer or shorter rods will work too, as long as they're at least 4" (10cm). Be aware that ferrite is brittle, so if the seller doesn't package it well, it may break in transit (and you end up with twice as many rods as you ordered, but half as long). If this happens, you can try to get a refund, but it is a hassle and takes time. I have not found these rods in the large electronics shops, so you may have to rely on something like eBay, Amazon, or AliExpress for it. Search for "ferrite rod," "ferrite bar," or "loopstick." I have had good experiences with An Ant Store on AliExpress.

VARIABLE CAPACITOR

The variable capacitor is another component not easily found from the largest vendors in the United States. The type you need is 223P. It has sections of 60pF and 140pF in the highest-capacitance setting, giving 200pF in total. Make sure you don't get the similar-looking 223F, which has a much lower maximum capacitance. A plastic tuning wheel is generally sold with variable capacitors and is almost essential. Using pliers to rotate the shaft of a variable capacitor is not much fun.

You may find variable capacitors listed as tuning capacitors. So long as the component has the 223P part number, it will be the same thing.

CONNECTION BLOCK

You will need a European-style connection block with 12 pairs of terminals spaced ⁵⁄₁₆" (8mm). (See Figure 1-25.) I'm specific about the terminal spacing because the block needs to fit the connectors of the variable capacitor. You will cut the block into sections of three, four, and five terminals, as shown in Figure 1-26.

SEMICONDUCTORS

There are three types of semiconductors you will need for the experiments in this book:

- BAT48 Schottky diode: Not all Schottky diodes work equally well, so don't substitute another one. It is available either with a reddish-brown glass body like regular small diodes, or with a blue body. I chose blue ones so that I could more easily distinguish them from regular silicon diodes, but either kind will work.

- 2N3904 NPN transistor: A common and cheap general-purpose transistor, bipolar NPN. Any manufacturer will work.
- LEDs: You need at least two LEDs—one red, and one of any color, 5mm or 3mm.

INTEGRATED CIRCUITS

For all of the following integrated circuits, get the through-hole version, which fits in a breadboard, not a surface-mounted variant. The chip-package variant you need is called DIP or PDIP. Avoid anything marked "SOIC," "SSOP," or "TSSOP," since these are surface mounted. These are the ones to look for:

- 7555 or ICL555: These are CMOS versions of the 555 timer chip. You need two of the CMOS versions.
- LM386 audio amplifier: In Experiment 10, the ones manufactured by Texas Instruments (previously National Semiconductor) appear to perform the best. There are three versions: LM386N-1, LM386N-3, and LM386N-4, and they appear to perform equally well.
- 4030B or 4070B logic chip quad two-input XOR gates: These have four XOR gates with two inputs each. 4011B, with four NAND gates, also works.

SLIDE SWITCH

Our experiments generally use a slide switch, or SPDT (single pole, double throw) as a power switch. The SS12D00 is a good option, but any switch will work as long as the pin spacing is 0.1" so that the pins fit adjacent breadboard rows.

PUSHBUTTON

You will need a momentary pushbutton, also known as a tactile switch, such as the SKRGAED010, the TE Connectivity FSM2JART, the Panasonic EVQ-PV205K, or the Mountain Switch 101-TS7311T1601-EV. It should have two pins spaced 0.2" apart to fit on the breadboard.

SPEAKER

The speaker should have a diameter of 2"–4" and an impedance of 8 ohms. The sound will be much better if it's mounted in a box of some kind. (See Experiment 1.)

BREADBOARD

This may be described as a breadboard or a solderless breadboard. You need one with 830 contact points and dual buses for supply voltage on either side of the board. A common model is MB-102.

I recommend getting at least two so you can have a transmitter and a receiver at the same time. A half-size breadboard (with about 400 contact points) is sufficient for some of the circuits—for example, those in Experiments 7 and 9. Note that on some breadboards, the power buses are interrupted at the center of the board. If so, the red and blue lines along the power buses are also interrupted at this point. If you have one of these breadboards, you need to bridge the gaps in the power buses with jumper wire.

HOOKUP WIRE

For making jumper wires to connect breadboard rows and for wiring components to the breadboard, you need hookup wire. The wire needs to be solid core (not stranded), thickness AWG 22 (to fit the breadboard well). You need four colors: preferably, red, green, yellow, and blue or black. It's convenient to get a set of several colors on spools in a box, and for breadboarding, such a set will last a long time. For winding coils and making antennas, you need considerably more wire. It may be easier to use thinner wire (down to AWG 26); however, I found that AWG 22 works for that as well. The total amount of wire used in our experiments is 105 feet, but there are possibilities for reuse, and it's possible to splice pieces for the coils. For coils and antennas, the wire color doesn't matter.

JUMPER WIRES

For general breadboarding, I recommend that you cut your own jumper wires from hookup-wire spools. (See above.) Cut them to the right length so they lie flat against the breadboard; otherwise, it's very hard to see if you have wired correctly. For the specific task of connecting the LCD module to the breadboard, I recommend using premade jumper wires. You need four wires with plugs at each end, of the same length, about 6", and they should be male-to-female, or MF, to fit in the breadboard and on the pins of the LCD module. Ideally, they should be red, black, green, and yellow.

HIGH-IMPEDANCE EARPHONE

A high-impedance earphone can be used with several projects in the book and is essential for the radio in Experiment 1—a normal, low-impedance earphone used for music players will not work there. High-impedance earphones are not common anymore but are still sold, often for crystal radio. (See Figure 1-34.) Unfortunately, the quality is questionable. (ProTechTrader.com sells a version with soldered connections, which is more reliable, and is recognized by having black wires instead of the regular beige ones.) A cheaper and more certain alternative, which is also easier to obtain, is a passive piezo buzzer.

Passive here means that the component doesn't generate its own waveforms but will create sounds according to the electric signal fed to it, which is exactly what is needed here. A high-impedance earphone contains a similar piezo-electric membrane; the difference is that the earphone has a plastic shell that fits the ear.

RASPBERRY PI PICO MICROCONTROLLER
The Raspberry Pi Pico microcontroller is used in Experiments 4 through 9. The experiments have been tested with an original Raspberry Pi Pico microcontroller. There are many variants with the same RP2040 microcontroller chip, and I cannot guarantee they are all compatible.

The Raspberry Pi Pico H has header pins soldered on. The header pins are necessary for connecting the Pico to the breadboard, so either get the H model or be prepared to solder them yourself. I have not tested the Wi-Fi version (indicated with a *W* at the end of the name).

If you have two Picos, you can use one to transmit remote control signals and another to receive them. If you have only one, you need to buy a transmitter button or a remote control instead and use your one Pico as a receiver.

LCD SCREEN
LCD screens with two rows of 16 characters are described as 1602 LCDs. Importantly, you need an LCD with the HD44780 controller (or compatible) and with an I2C interface, which makes it possible to control the display using only four wires from the Pico. The I2C functionality is generally provided by a separate, small circuit board mounted on the back of the LCD module. There are also similar LCD modules which have I2C functionality without the additional board mounted on the back. Search for "LCD screen 1602 with I2C." If you happen to have a larger LCD module with the same HD44780 controller and I2C, it will probably work (and you can adapt the sketches in the book to use the full display area if you want).

SI4703 FM RECEIVER MODULE
Use a SparkFun WRL-12938, or a clone so long as it uses the Si4703 FM receiver chip. If you don't have soldering equipment or want to avoid soldering, get a module with the pin headers already soldered on.

There are two versions of this module with the positive and negative power pins in different positions, so check the markings on the circuit board for the correct wiring.

433MHZ REMOTE CONTROL MODULES

I recommend the WPI469 433MHz Wireless Module Set (consisting of one transmitter and one receiver; see Figures 9-2 and 9-3). It's sold under the brand names Whadda, Velleman, and Pimoroni from sources including Jameco, RobotShop, AliExpress (search for "433 open-smart" and look for a blue circuit board), eBay, and Digikey. Pay attention to the frequency, as there is a similar module using the 315MHz frequency.

These modules (or the clones) are unique in that they have an antenna conveniently integrated as a trace on the circuit board. If you get some other remote control modules, they will likely work in the experiment, but you will have to solder on an antenna of the correct length. (See Experiment 9 for details.)

There is, however, one type of remote control receiver I have to recommend against: the cheapest variant, often called a supergenerative receiver. If they are sold as a pair, the transmitter in the pair is marked FS1000A, and the circuit boards are usually green. This receiver is considerably less sensitive than all the others.

If you have only one Pico board, you need one of the following for Experiment 9:

- A remote control operating on 433MHz with one or more buttons. (See Figure 9-1.) Make sure it supports protocol EV1527 or PT2262. Search for "433MHz EV1527" on eBay or Amazon.
- A 433MHz doorbell button that supports protocol EV1527 or PT2662. (The left-most device in Figure 9-1 is an example.)

MAKE: RADIO PARTS LIST

Chapter	1	2	3	4	5	6	7	8	9	10	11	12
RESISTORS												
22		1										1
100	1	2			1	1			1	1	2	2
330	1	1							1			1
1K			3			3	1		1			1
2.2K	2	4					2					1
4.7K			2							1	1	
6.8K		2										
10K	2	4	1			2			1			
47K		5										3
TRIMMER												
10K		1			1	1				1		1
500K	1											
CERAMIC CAPACITOR												
33pF												3
68pF												1
100pF	1	1					1					
220pF						1				1	1	1
470pF											4	1
1nF												3
2.2nF			1									
4.7nF		1										
10nF	3	8					1					
47nF		1	1			1						1
0.1uF	1	4	1						2		1	2
1uF			1		1	1	1	1				
ELECTROLYTIC CAPACITOR												
10uF		2	1			1				2		2
100uF	1	1							2	2	1	
470uF		2						1				2
INDUCTOR												
10uH												2
22uH			1								1	
1000uH										1		

Minimum	Total	Remarks
		Must be 5% tolerance, 4-band color code
1	1	Example: Yageo CFR-25JR-52-22R
2	3	Example: Yageo CFR-25JR-52-100R
1	1	Example: Yageo CFR-25JR-52-330R
3	4	Example: Yageo CFR-25JR-52-1K
4	4	Example: Yageo CFR-25JR-52-2K2
2	2	Example: Yageo CFR-25JR-52-4K7
2	2	Example: Yageo CFR-25JR-52-6K8
4	6	Example: Yageo CFR-25JR-52-10K
5	5	Example: Yageo CFR-25JR-52-47K
		Must fit breadboard
1	2	Example: Bourns 3362P-1-103
1	1	Example: Bourns 3362P-1-504
		Avoid Y5V type (see Appendix A); prefer ones with large, readable markings
3	3	
1	1	
1	1	
1	1	
4	4	
3	3	
1	1	
1	1	
8	8	
1	2	
4	5	
1	2	
2	3	Example: WCAP-ATG5 10uF 16V 20%
2	2	Example: WCAP-ATG5 100uF 16V 20%
2	2	Example: WCAP-ATG5 470uF 16V 20%
2	2	Example: Bourns 78F100J-TR-RC
1	1	Example: Bourns 78F220J-RC
1	1	Example: Bourns 78F102J-TR-RC

MAKE: RADIO PARTS LIST

Chapter	1	2	3	4	5	6	7	8	9	10	11	12	
IC													
7555	2	2					1						
LM386		1								1		1	
4030B/4070B/4011B											1		
MODULES													
Raspberry Pi Pico H				1	1	1	1	1	1(2)				
LCD				1	1	1	1	1					
SparkFun WRL-12938 Si4703 FM board								1					
433MHz remote pair									1				
433MHz button or remote*									1				
SEMICONDUCTORS													
BAT48	1			2	2	2	4	2				1	
Red LED**		1							2				
2N3904		3	1			1			1			1	
MISC													
SPDT slide switch	1	1								1	1	1	
pushbutton				1	1	1	1	3	6				
9V battery	1	1	1							1	1	1	
ferrite rod	1	1								1	[1]		
tuning cap	1	1	1							1		2	
speaker	1	1						1	1			1	
audio cable		1				1		1					
audio adapter		1				1		1		1	1	1	
breadboard	1	1	1	1	1	1	1	1	1(2)	1	1	1	
Micro USB				1	1	1	1	1	1(2)				
MF jumper				4	4	4	4	4					
high-impedance earphone	1												
Euro connection block	1	1	1							1		1	
9V battery connector	1	1	1							1	1	1	
22-gauge wire													
22-gauge wire	40										13		
26-gauge wire	6	15				30							

* Experiment 9 needs either a 433MHz button or remote (EV1527 or PT2262 protocol) *or* two Picos and two breadboards.
** One LED needs to be red. The other can be any color.

	Minimum	Total	Remarks
	2	2	Must be CMOS; do not substitute TTL part
	1	1	Prefer National Semiconductor or Texas Instruments
	1	1	
	1	2	
	1	1	1602 with I2C
	1	1	
	1	1	WPI469 or Open-SMART clone
	1	1	EV1527 or PT2262 protocol
	4	4	Prefer blue body in order to not mix with the common 1N4148
	2	2	
	3	4	
	1	1	Leads spaced 0.1" to fit breadboard
	6	6	Leads spaced 0.2" to fit breadboard
	1	2	
	1	1	10mm x 160mm (0.4" x 6.3")
	2	2	
	1	1	
	1	1	With 1/8" plugs
	1	1	1/8" audio jack to screw terminals
	1	2	
	1	2	
	4	4	
	1	1	Can substitute piezo buzzer; see Experiment 1 and Appendix A
	1	1	Terminals spaced 5/16" (8mm); may be described as H-type, 3A
	1	2	
	3 feet	3 feet	In each of the colors red, black, yellow, green; solid core.
	40 feet	40 feet	Any color; solid core.
	30 feet	30 feet	Any color, for winding coils. Solid core; thicker wire, up to 22-gauge, also works—see Appendix A.

APPENDIX B: SOURCES FOR KITS, TOOLS, AND COMPONENTS

KITS

ProTechTrader (www.protechtrader.com)

COMPONENT SUPPLIERS

There are three large, professional component suppliers in the United States that will also sell to small-scale customers: Mouser (www.mouser.com), DigiKey (www.digikey.com), and Newark (www.newark.com).

These are reliable—you will get exactly what you order, and their websites offer a datasheet for every component so you can check the properties. There are multiple options, so, for example, when you shop for a 10K resistor, you will see hundreds of variants by different manufacturers, mounted or surface mounted, and so on. The downside of this is that it can be hard to choose. You have to learn to use the search functions on the store web pages. The suppliers sell most components in single quantities but often offer much cheaper prices if you buy in bulk. For example, there is almost no point in buying a single resistor; I tend to buy a hundred for each value I need. Shipping is not free (or there may be a minimum order amount for free shipping), so try to group as much in a single order as you can.

In Europe, I have had good experiences with Mouser and Reichelt (www.reichelt.de).

EBAY AND ALIEXPRESS

Some components for this book cannot be found in the large component shops (in particular, the variable capacitors and the ferrite rod; see Appendix A). And sometimes you can find better deals from online suppliers, which are just brokers for a huge number of different sellers. The two best-known examples are probably eBay (www.ebay.com or regional variants) and AliExpress (www.aliexpress.com).

In my own experience, packages from AliExpress always arrive, usually in less than two weeks (but occasionally much longer). The quality is generally fine, but one potential issue is that there may not be detailed information such as an exact part number or a datasheet.

I generally don't recommend specific sellers since my experience with any single one is limited and I don't know how long they will stay operational. For the ferrite rods (which are brittle), however, I mentioned An Ant Store on AliExpress, since they shipped them with serious padding.

It's worth shopping around for these components in particular:

- Ferrite rods
- Variable capacitors
- FM radio modules
- 433MHz transmitter and receiver modules
- 433MHz doorbell buttons or remote control transmitters
- I2C LCD screens
- Breadboards (buy several)
- Euro connection blocks
- Audio jack to screw adapters
- Wire strippers (for example, YTH-5023)

HOBBYIST STORES

Hobbyist stores may be a good source for breadboards, tools, hookup wire, and boards such as the Raspberry Pi Pico and the FM receiver.

SparkFun (www.sparkfun.com) designs and sells lots of modules, such as the FM-receiver module we use for this book.

Adafruit (www.adafruit.com) sells tools and modules.

If you're in Europe, Kiwi Electronics (www.kiwi-electronics.com) is another good resource.

Color values

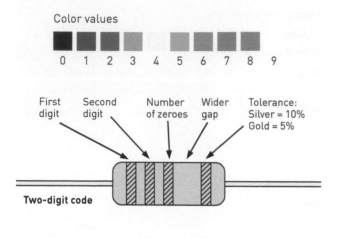

0 1 2 3 4 5 6 7 8 9

First digit   Second digit   Number of zeroes   Wider gap   Tolerance: Silver = 10% Gold = 5%

Two-digit code

First digit   Second digit   Third digit   Number of zeroes   Wider gap   Tolerance (various colors)

Three-digit code

C-1 *Resistor color code.*

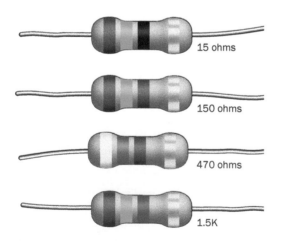

15 ohms

150 ohms

470 ohms

1.5K

C-2 *Examples of resistor color codes.*

APPENDIX C: RESISTOR AND CAPACITOR MARKINGS

This appendix explains how to read the markings on resistors and capacitors.

RESISTORS

The resistors used in this book are marked with colored bands. The colors are a code for describing the resistor value in ohms and the tolerance. Using colored bands to show the value allows you to read the value from any angle, even if the resistor is soldered onto a circuit board.

To build a circuit, you need to be able to tell the resistors apart. The breadboard diagrams throughout this book show the resistors with their colored stripes, which will help you to find the correct component. It is also worth learning to decode the color bands yourself.

Resistors are usually marked with four or five colored bands. The four-band version is easier to read. Each band denotes a digit, as shown in Figure **C-1**. To help you remember the values, note that the colors from red (2) to violet (7) are ordered as in the rainbow.

This is the procedure: Modern four-band resistors almost always have a gold band, denoting 5% tolerance. Turn the resistor so the gold band is to the right. Start reading the bands from the left. To translate the digits to a resistance value, take the first two digits followed by the number of zeroes given by the third digit. Practice on the resistors in Figure **C-2**.

Five-band resistors use the same system, except that there are three digit bands and the fourth band gives the number of zeroes. The fifth band gives the tolerance and is usually brown (meaning 1%). This makes it harder to see which way to start reading the code, although there is generally a larger gap between the tolerance band and the others. I strongly recommend getting four-band resistors so you can practice quick reading.

This book uses the following resistors:

Color Sequence	Numbering	Resistance
red red black gold	2-2-0-5%	22
brown black brown gold	1-0-1-5%	100
orange orange brown gold	3-3-1-5%	330
brown black red gold	1-0-2-5%	1K
red red red gold	2-2-2-5%	2.2K
yellow violet red gold	4-7-2-5%	4.7K
blue gray red gold	6-8-2-5%	6.8K
brown black orange gold	1-0-3-5%	10K
yellow violet orange gold	4-7-3-5%	47K

Resistance is measured in ohms. In some books, the Greek letter Ω is used to represent this unit, but many schematics omit the symbol, and I follow that convention in this book. Larger resistors use the units kilohm (1,000 ohms) and megohm (1,000,000 ohms). In schematics, these units are written as *K* and *M*. So, *47K* means "47,000" ohms. Sometimes, but not in this book, you may see *3K3* instead of *3.3K*. This saves one character and avoids relying on a decimal dot, which may be hard to see if the text is small.

Resistors are manufactured with standard values, called E-series. In this book, I've selected resistors from the series called E6, in which the first two digits of the resistance value can be 10, 15, 22, 33, 47, or 68. Using this series means that there is a limited number of color codes, and after a while, you will learn to quickly recognize the color codes without referring to the table for each individual color band. Anyway, it is a good habit to check resistor values with a multimeter to be sure.

CAPACITORS

Capacitance is measured in farads, using the letter *F*. As a farad is a very large unit, practical capacitors are measured in picofarads (pF), nanofarads (nF), microfarads (µF), or millifarads (mF):

- 1,000pF = 1nF
- 1,000nF = 1µF
- 1,000µF = 1mF
- 1,000mF = 1F

(The millifarad is rarely used.)

This book uses two kinds of capacitors: electrolytic capacitors for values above 1µF and ceramic capacitors for 1µF and below.

Electrolytic capacitors are clearly marked with the capacitance in µF and a maximum voltage. Electrolytic capacitors are polarized: They have a positive and negative lead and must be connected the correct way. The negative lead is shorter and marked with a stripe on the capacitor body, sometimes also with a minus sign. In schematics, electrolytic capacitors have a plus sign marking the positive plate (to remind you that it's polarity-sensitive) and the negative plate is curved.

Ceramic capacitors are physically quite small, so there is limited space for markings. They are generally marked with a compact code, usually just three digits. These digits give the capacitance in picofarads. As in the resistor color code, the two first digits denote the first digits of the capacitance. The third digit gives the number of zeroes.

This book uses the following ceramic-capacitor values:

Capacitance Code	Capacitance Value
330	33pF
680	68pF
101	100pF

Capacitance Code	Capacitance Value
221	220pF
471	470pF
102	1,000pF = 1nF
222	2,200pF = 2.2nF
472	4,700pF = 4.7nF
103	10,000pF = 10nF
473	47,000pF = 47nF
104	100,000pF = 0.1µF
105	1,000,000pF = 1µF

Note especially the first two rows, where the final zero can be confusing: 330 means 33pF (that is, no zeroes) and not 330pF.

INDUCTORS

In some experiments, you make inductors yourself by winding hookup wire on a ferrite rod or on a nonmagnetic object. You can also buy ready-made inductors as components, such as for Experiments 3, 10, 11, and 12. The kind I suggest look like resistors and use the same (four-band) color code for the inductance value in µH. Inductance is measured in henries, abbreviated *H*. Like the farad, a henry is a large amount of inductance, so practical inductors have values in microhenries (µH) or millihenries (mH). The following ready-made inductors are used in this book:

Color Sequence	Numbering	Resistance
brown black black gold	1-0-0-5%	10µH
red red black gold	2-2-0-5%	22µH
brown black red gold	1-0-2-5%	1,000µH = 1mH

APPENDIX D: UNDERSTANDING OSCILLOSCOPES

This final appendix explains the fundamental features of an *oscilloscope*, which can be a very valuable tool for anyone with an active interest in electronics.

This gadget is most often used to display a graphical image of a voltage that may change rapidly inside a circuit. Your multimeter can't display rapid changes, because it requires a moment to "settle" when you are measuring a DC voltage. The meter has an AC setting, in which it tells you the value of an alternating voltage, but only if the voltage fluctuates regularly and repeatedly basis, within a limited range of frequencies. For example, a Fluke 179 is considered a high-end meter but cannot measure frequencies above 10kHz. Other meters have similar limitations; an oscilloscope does not.

What are some specific situations where you would use an oscilloscope? Here are a few examples:

- To see the shape of an oscillating signal. The scope will show whether it's a square wave, a triangle wave, a sine wave, an audio wave, or something else.
- To see if a DC power supply is delivering a steady voltage, is free from fluctuations, and remains accurate at various loads.
- To see if the output from a component contains a transient voltage spike.
- To see rapidly changing phenomena such as the displacement current through a capacitor, when voltage rises or falls suddenly.

An oscilloscope screen is like an MRI image that reveals the inside of the human body, except that the screen shows you the inner workings of a circuit. However, there are some things that an oscilloscope won't do. In particular, it won't measure current, resistance, or capacitance.

TYPES OF OSCILLOSCOPES

Decades ago, all oscilloscopes were very expensive, heavy, *analog* devices using monochrome cathode-ray tubes for their display. You may think you can save some money by buying one of these old beasts secondhand, but its performance will be inferior to modern equipment that is quite modestly priced.

A *digital* oscilloscope converts a varying voltage input into digital samples and displays them as patterns of pixels on an LCD screen. Digital scopes are either benchtop or handheld devices. Figure **D-1** shows a very low-cost, handheld product, Figure **D-2** shows a benchtop model, and Figure **D-3** shows a tablet version that has a touch screen. You can also buy a multimeter that has a limited range of oscilloscope functions, such as the example in Figure **D-4**.

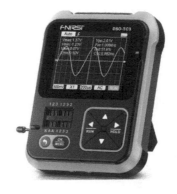

D-1 The FNIRSI DSO-TC3 is a very low-priced, palm-size oscilloscope with limited features.

In all of these examples, the screen has limited resolution. Most will allow you to connect the output to a computer via a USB cable to display a more detailed picture—but if that is your intention, you can probably save money by choosing a **USB oscilloscope** (also sometimes known as a **PC oscilloscope**). This doesn't have its own screen. It just consists of a little plastic box, such as the one shown in Figure **D-5**, with ports for input leads and a USB output. Software provided by the oscilloscope manufacturer runs on your computer, interprets the signal, and displays it on the computer screen. In this way, the oscilloscope offloads the chore of maintaining a display, which reduces its cost while also generating a relatively high-resolution image.

D-2 The Hantek DSO2C10 is a popular desktop oscilloscope.

D-3 The FNIRSI 1013D Plus is a modestly priced, tablet-size digital oscilloscope with a touchscreen.

If you are learning electronics, a USB oscilloscope makes so much sense, I strongly suggest that you consider this type. The PicoScope 2204A is currently available for well under $200 in the United States (and often found secondhand on eBay). I used it to generate all the oscilloscope readings shown in this book. Other USB scopes, from manufacturers such as Hantek, work in much the same way and may cost a little less while omitting some features.

D-4 The Liumy LM2020 is a multimeter that has some limited oscilloscope capabilities built in.

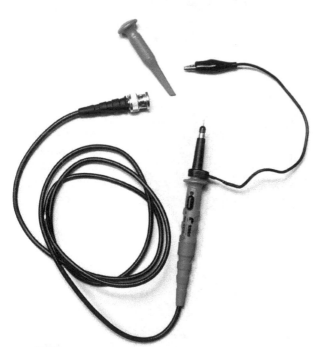

D-5 *The PicoScope 2204A is very modestly priced but provides a full set of features in software. Two probes plug into the coaxial sockets at this end of the box, while there is a USB port at the opposite end.*

D-6 *An oscilloscope probe. The alligator clip attaches to any ground wire in your circuit.*

Since a USB oscilloscope has to work with a computer, you may be wondering how you're going to satisfy that requirement. How powerful does the computer have to be? And what if your existing computer isn't located near your electronics bench?

One option is to buy the cheapest possible secondhand laptop, which you can keep permanently on your workbench, where you can also use it for tasks such as writing sketches for your Pico microcontroller. (Note that the PicoScope is not made by the same company as the Pico microcontroller. In both cases, the term *pico* is based on the Spanish word meaning "small.") Almost any laptop capable of running at least Windows 10 will be adequate. If you prefer a Mac, make sure that the version of macOS you're running will support the oscilloscope software.

If you are thrifty, and you shop around, you may be able to put together a used laptop and a used USB oscilloscope for less than the cost of a new benchtop oscilloscope, and the display will be in higher resolution.

SETUP

You need to maintain a clear mental distinction between the hardware of a USB oscilloscope and the software that runs on your computer to render the image. The instructions for the hardware will be very brief, as you simply plug in the test leads (usually two of them, such as the one shown in Figure **D-6**) and a USB cable to your computer. The scope often takes its power through the USB connection, so there may not be a separate power supply. You may have to supply your own USB cable to connect the scope with the computer, but it is often the same as a printer cable.

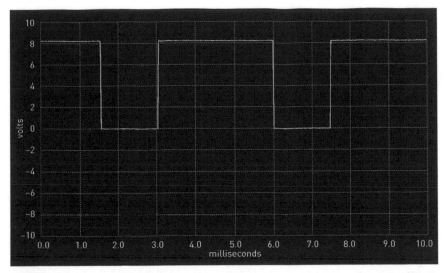

D-7 *Displaying the output from a 555 timer running in astable mode.*

The software may have its own separate manual available for download. Note that a manufacturer may use the same software for several different versions of the hardware. For instance, PicoScope software version 6 will work with almost all PicoScope models, including the 2000, 3000, 4000, 5000, and 6000 series.

BASIC CONCEPTS

When the oscilloscope software is running on your computer, you can use either of the probes to measure voltage at any point in a circuit. Each probe has a ground wire attached, which you should connect with the negative ground in your circuit, although the scope is still likely to show a signal if you don't.

The maximum voltage that a low-cost USB oscilloscope can measure is often +/-20V. The probes have such a high resistance, they will protect the scope from voltages outside of that range. The resistance also means that the probes will interfere as little as possible with the functioning of your circuit.

Figure **D-7** shows a square wave generated at the output pin of a 555 timer chip (not a 7555; this is the old, original TTL version of the chip). The horizontal line at the bottom of the image shows time in milliseconds, while

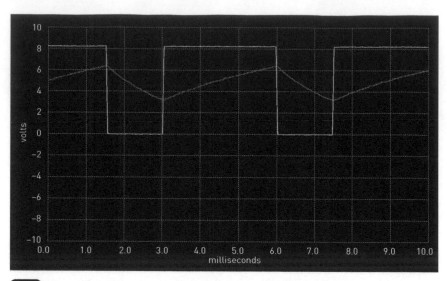

D-8 *The charging and discharging of the timing capacitor controlling a 555 timer has been added in red.*

the vertical line at the left side of the image shows volts. You read the graph from left to right.

You can change the scales to match the voltage range and frequency of the signal, or you can run the oscilloscope in auto mode, so that it chooses the scales for you. Since the 555 creates a regularly repeating signal, you can tell the scope to resample it automatically at intervals to reveal any unexpected changes. Alternatively, you can tell the scope to take a snapshot for you and store it. Another option is to set a trigger point so that when the voltage rises or falls past that point, a snapshot is triggered automatically.

In Figure **D-8**, the second probe for the oscilloscope has been attached to Pin 6 of the 555 timer, which is connected with a timing capacitor. Now, you can see how the voltage on the capacitor varies while the square wave output switches on and off.

Each probe generates a different-colored graph line on the screen, and the setup options for the software will enable you to change those colors if you wish. You can also change the look of the screen—the background color, the grid lines, and other features.

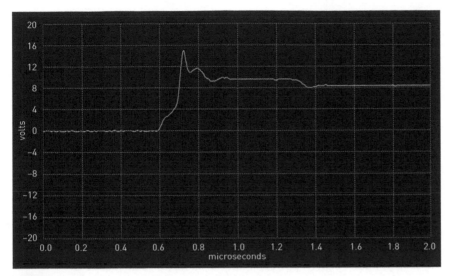

D-9 *When the timer output is viewed in microseconds, its imperfections are revealed.*

Now, let's suppose you change the scale on the screen so that you zoom in on the signal, as in Figure **D-9**. The voltage represented by each interval on the vertical axis has doubled, while the elapsed time represented by each interval on the horizontal axis has decreased by a factor of 5,000.

Maybe you have read in the past that a 555 timer generates a voltage spike; well, now you can see it for yourself. The 555 output that looked like a nice, sharp square wave turns out to reach almost 15V before it gradually flattens out. Although the disturbance is brief, it can confuse other components in a circuit. (That's why we've used 7555 timers in this book. They have a much more stable output.)

TERMINOLOGY
The graph line on the screen is known as a **waveform**, or a **trace**. The horizontal scale is known as the **x-axis**, while the vertical scale is known as the **y-axis**. The x-axis usually displays the **collection time**, while the y-axis shows the **input range** in volts or fractions of a volt.

The two probes are identified by the software as **Channel A** and **Channel B**. Somewhere on the screen you should see a display of the frequency of an oscillating signal, if the oscillations are occurring at regular intervals. How high a frequency can an oscilloscope measure? At least 10MHz. To make sure of this, you can set the **sampling rate**, which should generally be at least five times the frequency.

You will be able to save multiple snapshots in a graphical format such as JPEG, PNG, or BMP. The most desirable option is probably PNG, as it creates a small file but is lossless, so there will be no visual artifacts. You can also save the mathematical values that generated a waveform so that you can reload it into the software for analysis later. And you should be offered an option such as .csv, which will enable you to save data in a format that a spreadsheet program can open.

Already, you can see how powerful and versatile an oscilloscope is, compared with your humble multimeter. I have only summarized some basic features here; the software manual for any USB scope that you buy should describe a wealth of capabilities to help you capture and analyze data.

Radio consists of signals that usually change rapidly. You will obtain a much clearer idea of what is really happening if you view them with an oscilloscope.

INDEX

volatile 125
VSYS 83

W

waveform 6, 243
wavelength 1, 7
Wheeler's approximation 23
WiFi 214
Wire library 89

X

XOR gate 184, 223

ACKNOWLEDGMENTS

I am grateful to everyone who contributed to this book.

Charles Platt, besides drawing the figures and writing the foreword, was of great help with the whole project and helped me structure the book for learning by discovery.

Copy editor Sophia Smith, with her sharp eye for detail, helped keep the book consistent and correct, both in language and in terminology.

Designer Juliann Brown did the layout and developed the clear and appealing appearance of the book, creating a far more beautiful finished work than I could have imagined while writing.

Editor Kevin Toyama navigated the book-production process and patiently helped bring the book to the finish line. He offered valuable encouragement and enthusiasm along the way, helping with keeping the experiments approachable, test-building some of the circuits himself, and sharing his experience.

I have learned a great deal from working with all of you, and you have turned this project into a book beyond anything I could expect.

Andrea Klettke, Bernardo Dias, Ludovica Guarneri, Mark Botham, Thomas Bauer, Tom Hoekstra, and Vincent Mondamert helped out greatly with test-building and improving the instructions. It was enlightening and a lot of fun to build together and see the projects in this book come to life. Finally, I am grateful to Johanna Grönqvist for organizing the test-building event and for her support throughout the project.

ABOUT THE AUTHORS

Minnie Middelberg, CWI

FREDRIK JANSSON enjoys tinkering with electronics and is a licensed radio amateur. He works as a researcher on the physics of clouds at the Delft University of Technology and develops weather models that run on supercomputers. He has a PhD in Physics from Åbo Akademi University in Finland, and lives in Amsterdam.

CHARLES PLATT is a contributing editor and regular columnist for *Make:* magazine, where he writes about electronics and tools. Platt was a senior writer for *Wired* magazine, has written various computer books, and has been fascinated by electronics since he put together a telephone answering machine from a tape recorder and military-surplus relays at age 15. He lives in a Northern Arizona wilderness area, where he has his own workshop for prototype fabrication and the projects that he writes about for *Make:* magazine.